The Kinematics of Machinery

The Kinematics of Machinery

Justin Rosales

NY RESEARCH
PRESS

New York

Published by NY Research Press
118-35 Queens Blvd., Suite 400,
Forest Hills, NY 11375, USA
www.nyresearchpress.com

The Kinematics of Machinery
Justin Rosales

International Standard Book Number: 978-1-64725-431-5 (Hardback)

Cataloging-in-Publication Data

The kinematics of machinery / Justin Rosales.
  p. cm.
Includes bibliographical references and index.
ISBN 978-1-64725-431-5
1. Machinery, Kinematics of. 2. Machinery, Dynamics of.
3. Mechanical movements. I. Rosales, Justin.
TJ175 .K56 2023
621.811--dc23

Contents

Preface

The purpose of the book is to provide a glimpse into the dynamics and to present opinions and studies of some of the scientists engaged in the development of new ideas in the field from very different standpoints. This book will prove useful to students and researchers owing to its high content quality.

Kinematics refers to the study of motion in a system of bodies, which does not directly take into account the potential fields or forces influencing the motion. It uses numbers, words, equations, diagrams, and graphs to explain the motion of objects. Kinematic motion can be influenced by three different types of forces, which include gravity, contact force, and fluid pressure. In astrophysics, kinematics is used to explain the motion of celestial bodies and groups of such bodies. Kinematics of machines is the study of the relative motion of machine parts. It entails the investigation of displacement, acceleration, position, and velocity of machine parts. The information on these aspects of machines is useful for designing the mechanism of machine. This book covers in detail some existent theories and innovative concepts revolving around the kinematics of machinery. It aims to shed light on some of the unexplored aspects of this area of study and the recent researches in it. The book will provide comprehensive knowledge to the readers.

At the end, I would like to appreciate all the efforts made by the authors in completing their chapters professionally. I express my deepest gratitude to all of them for contributing to this book by sharing their valuable works. A special thanks to my family and friends for their constant support in this journey.

Justin Rosales

Link, Degrees of Freedom and Kinematic

1.1 Link

A single part (or an assembly of rigidly connected parts) of a machine, which is a resistant body having a relative motion with the other parts of machine is known as a Link or Element.

It is defined as a machine element having relative motion with respect to other parts of machine elements.

The link may consist of one or more resistant bodies. A link may be called as kinematic link or element. E.g.: Reciprocating linkage.

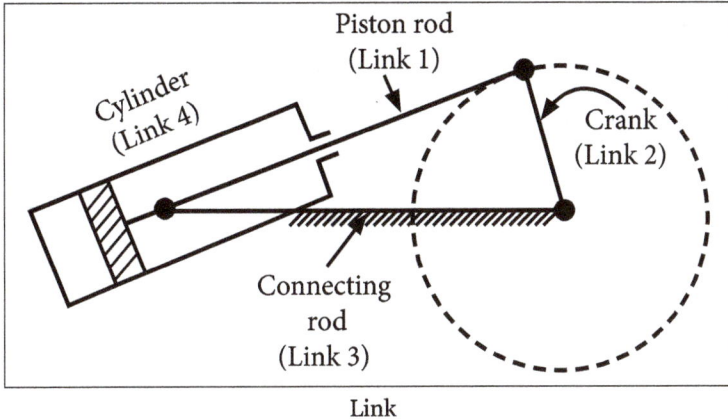

Link

Classification of Links

Depending upon the Flexible Property:

- Rigid Link: Rigid links are those links that does not undergo any change of shape when they transmit motion. In reality, no rigid links exist. But kinematic links whose deformations are very small are considered as rigid links. These links do not undergo significant deformation while transmitting motion.

  Examples: Connecting rod and crank pin in a steam engine, bed and spindle of a lathe do not have appreciable deflection and as such they can be termed as rigid links.

- Flexible Link: A flexible link undergoes partial deformation while transmitting motion. Its deformation does not affect its transmission effectiveness.

  Examples: Flexible links are belts (in belt drives) and chains (in chain drives).

- Fluid Link: A fluid link makes use of a fluid (liquid or gas) to transmit motion, by means of pressure. Fluid links undergo deformation while transmitting motion. This is one which is having a fluid in a receptacle and the motion is transmitted through the fluid by pressure or compression only.

  Examples: Pneumatic punching presses, hydraulic jacks and hydraulic brakes.

Depending upon the number of ends provided:

- Binary Link: Having two ends for turning pairs.

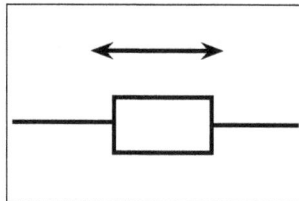

- Ternary Link: Having three ends for turning pairs.

- Quaternary Link: Having four ends for turning pairs.

1.1.1 Kinematic Pairs

When any two links or Elements are connected in such a way that their relative motion is completely or successfully constrained, they form a kinematic pair.

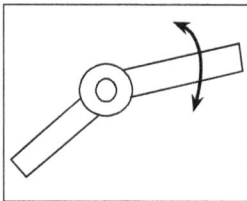

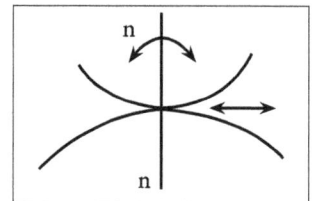

| (a) Turning pair | (b) Prismatic pair | (c) Higher pair |

Lower Pair and Higher Pair

Lower Pair: When the two elements of a pair have a surface contact when relative motion takes place and surface of one element slides over the surface of the other, the pair formed is known as lower pair.

Example: Sliding pair, turning pair and screw pair forms lower pair.

Higher Pair: When the two elements of a pair have a line (or) point contact when relative motion take places and the motion between the two elements is partly turning had partly sliding, then the pair is known as higher pair.

Example: Toothed gear, belt and rope drive, Roller bearing, cam and follower.

Translation and Rotation

Translation: A body has translation if it moves so that all straight lines in the body move to parallel positions.

Example: link 4 in the below figure:

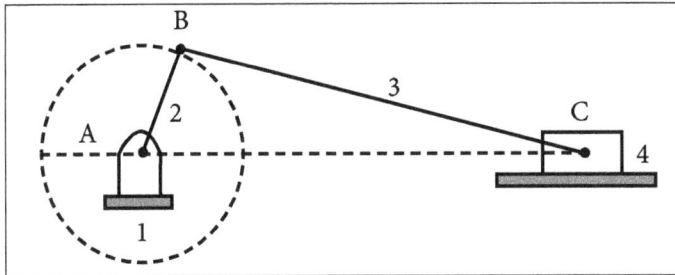

Higher Pair.

Rotations: In rotation, all points in a body remain at fixed distances from a line which is perpendicular to the plane of rotation. This line is the. Axis of rotation and points in the body describe circular paths about it.

Example: Link 2 in the above figure.

1.2   Degrees of Freedom

The number of input parameters which must be independently controlled in order to bring the mechanism into a useful engineering purpose.

Degrees of freedom of a pair is defined as the number of independent relative motions, both translational and rotational, a pair can have.

Degrees of freedom = 6 − no. of restraints.

To find the number of degrees of freedom for a plane mechanism we have an equation known as Gruebler's equation and is given by,

$$n = 3(l - 1) - 2j - h,$$

$$l = 3, d = 2, h = 1,$$

$$= 3(3 - 1) - 2(2) - 1,$$

$$n = 1$$

F = 0, results in a statically determinate structure.

If F > 0, results in a mechanism with 'F' degrees of freedom.

F < 0, results in a statically indeterminate structure.

Working of Kutzbach Criterion

The mobility of a mechanism is defined as the number of input parameters (usually pair variables) which must be controlled independently in order to bring the device into a particular position.

It is possible to express the number of degrees of freedom of a mechanism in terms of the number of links and the number of pair connections of a given type. This is known as number synthesis.

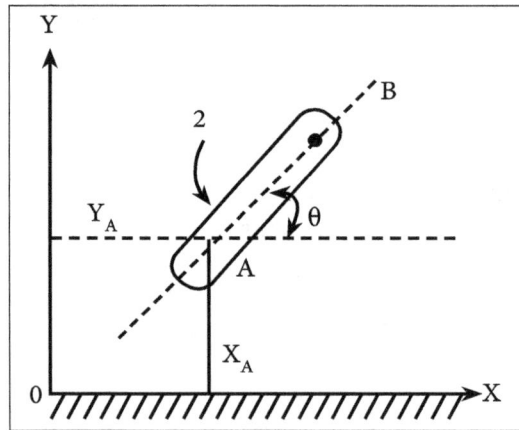

Let there are two links 1 and 2 in which link 1 is fixed, as shown in figure. The link 2 has a point A over it and translated by co-ordinates X_a and Y_a.

It can be written as A (x_a, Y_a).

A and B makes an angle θ with the fixed link 1 (OX). Link 2 specified by three variable (X_A, Y_A).

Let, Number of Links = 1.

So, Number of Movable links = (1 - 1) and total number of degree of freedom before they are connected to any other link = 3 (1 - 1).

If, J = Number of binary Joints or lower pair.

H = Number of higher pairs,

$n = 3(l - 1) - 2j - h$.

This equation is called Kutzbach criterion for the movability of a mechanism having plane mechanism.

1.2.1 Gruebler's Criterion

The degrees of freedom of a mechanism are the number of independent relative motions among the rigid bodies. For example, given below shows several cases of a rigid body constrained by different kinds of pairs.

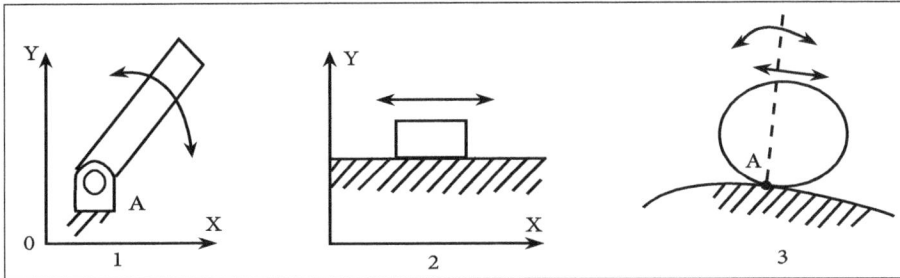

(a) Rigid bodies constrained by different kinds of planar pairs.

In the figure (1) a rigid body is constrained by a revolute pair which allows only rotational movement around an axis. It has one degree of freedom, turning around point A. The two lost degrees of freedom are translational movements along the x and y axes. The only way the rigid body can move is to rotate about the fixed point A.

In figure (2), a rigid body is constrained by a prismatic pair which allows only translational motion. In two dimensions, it has one degree of freedom, translating along the x axis. In this example, the body has lost the ability to rotate about any axis and it cannot move along the y axis.

In figure (3), a rigid body is constrained by a higher pair. It has two degrees of freedom: translating along the curved surface and turning about the instantaneous contact point.

In general, a rigid body in a plane has three degrees of freedom. Kinematic pairs are constraints on rigid bodies that reduce the degrees of freedom of a mechanism. The figure (a) shows the three kinds of pairs in planar mechanisms. There are number of the degrees of freedom. If we create a lower pair (1),(2), the degrees of freedom are reduced to 2. Similarly, if we create a higher pair (3) the degrees of freedom are reduced to 1.

Therefore, we can write the following equation:

$$F = 3(n-1) - 2l - h.$$

Where,

F = Total degrees of freedom in the mechanism.

n = Number of links (including the frame).

l = Number of lower pairs (one degree of freedom).

h = Number of higher pairs (two degrees of freedom).

Problem

The transom above the door is shown in the figure (a) below. The opening and closing mechanisms are shown in figure (b). Let us calculate its degree of freedom.

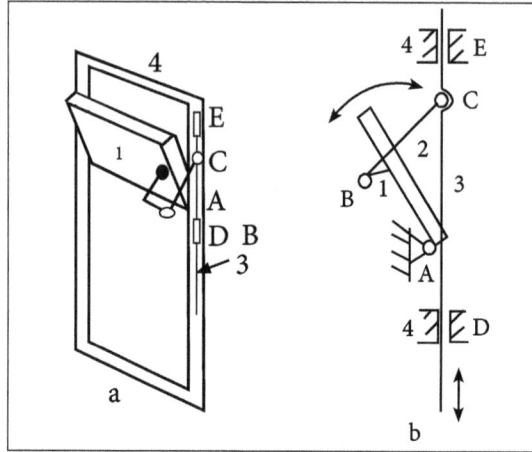

n = 4 (link 1,3,3 and frame 4), l = 4 (at A, B, C, D), h = 0

$$F = 3(4-1) - 2 \times 4 - 1 \times 0 = 0$$

Note: D and E function as a same prismatic pair, so they only count as one lower pair.

1.2.2 Kinematic Chain, Mechanism and Structure

Kinematic Chain

A kinematic chain is arrangement of kinematic pairs in such a way that each link forms a part of two pairs and the motion of each relative to other is definite and the last link is joined to the first link to transmit definite motion.

Example: Beam Engine and slider crank mechanism

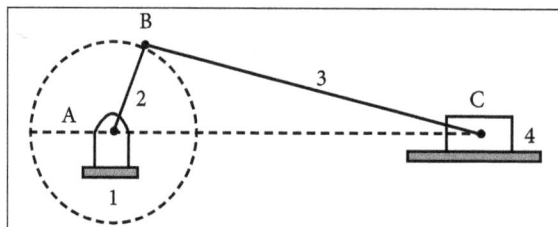

Slider Crank mechanism.

Mechanism and Structure

The degree of freedom of an assembly of links completely predicts its character. There are only three possibilities. If the DOF is positive, it will be a mechanism and the links will have relative motion. If the DOF is exactly zero, then it will be a structure and no motion is possible.

If the DOF is negative, then it is a pre-loaded structure, which means that no motion is possible and some stresses may also be present at the time of assembly. The figure shows examples of these three cases. One link is grounded in each case.

(a) Mechanism—DOF =+1   (b) Structure—DOF =0   (c) Freeloaded structure—DOF = -1

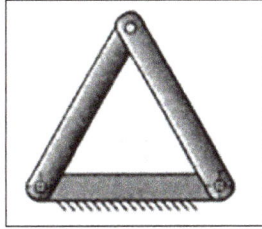

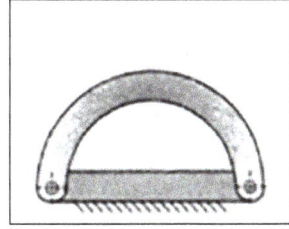

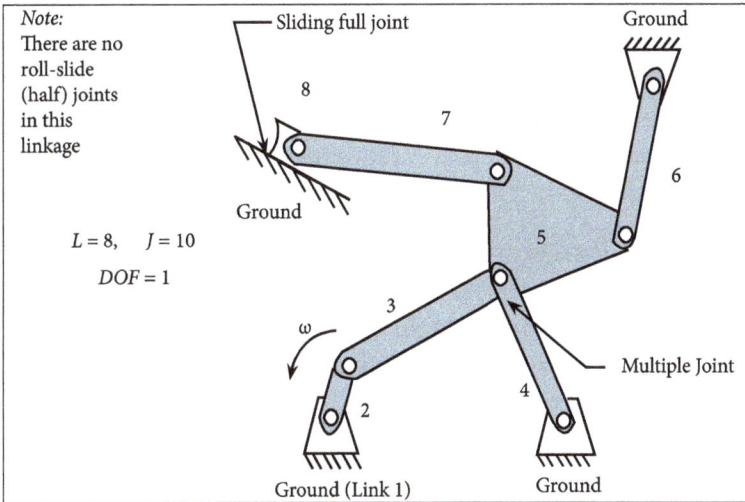

Linkage with full and multiple joint.

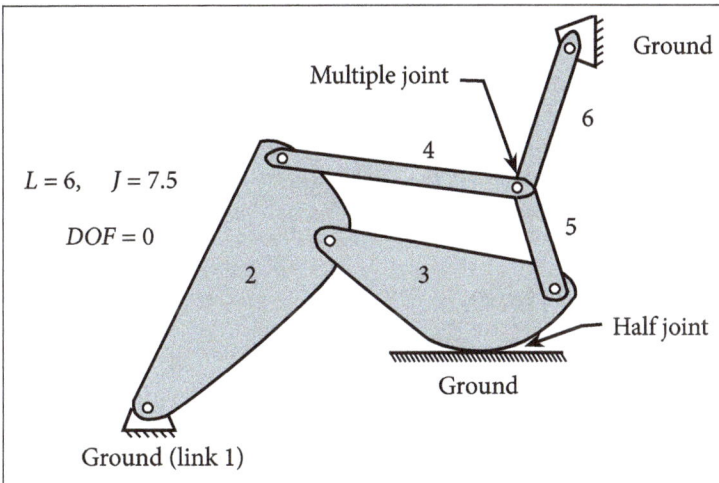

Linkage full, half and multiple joint.

The figure (a) shows four links joined by four full joints which from the Gruebler equation, gives one DOF. It will move and only one input is needed to give predictable results.

The figure (b) shows three links joined by three full joints. It has zero DOF and is thus a structure. Note that if the link lengths will allow connection, all three pins can be inserted into their respective pairs of link holes (nodes) without straining the structure, as a position can always be found to allow assembly. This is called exact constraint.

The figure(c) shows two links joined by two full joints. It has a DOF of minus one. Making it a preloaded structure. In order to insert the two pins without straining the links, the center distances of the holes in both links must be exactly the same. It is impossible to make two parts exactly the same. There will always be some manufacturing error, even if very small.

Thus we may have to force the second pin into place: creating some stress in the links. The structure will then be pre-loaded. It is similar to a situation in applied mechanics in the form of an indeterminate beam, one in which there were too many supports or constraints for the equations available. An indeterminate beam also has negative DOF, while a simply supported beam has zero DOF.

Both structures and pre-loaded structures are commonly encountered in engineering. In fact the true structure of zero DOF is rare in civil engineering practice. Most buildings, bridges and machine are pre-loaded structures, due to the use of welded and riveted joints rather than pin joints. Even simple structures like the chair we are sitting in are often pre-loaded. Since our concern here is with mechanisms, we will concentrate on devices with positive DOF only.

Difference between Mechanism and Structure

S. No.	Mechanism	Structure
1.	Mechanism transmits and modifies motion.	No relative motion exists between its members.
2.	A mechanism is the skeleton outline of the machine to produce definite motion between various links.	It does not convert the available energy into work.
3.	Example: Clock work, Typewriter.	Example: Shaper and Lathe.

1.2.3 Mobility of Mechanism

The mobility of a mechanism is defined as the number of input parameters (usually pair variables) which must be controlled independently in order to bring the device into a particular position.

A kinematic chain is a group of links either joined together or arranged in a manner that permits them to move relative to one another. If the links are connected in such a way that no motion is possible, it results in a locked chain or structure.

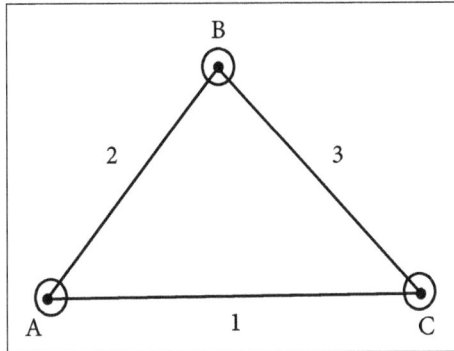

(a) Locked chain or structure.

The number of degrees of freedom of a mechanism is also called the mobility of the device. The mobility is the number of input parameters (usually pair variables) that must be independently controlled to bring the device into a particular position. The Kutzbach criterion, which is similar to Grubler's equation, calculates the mobility.

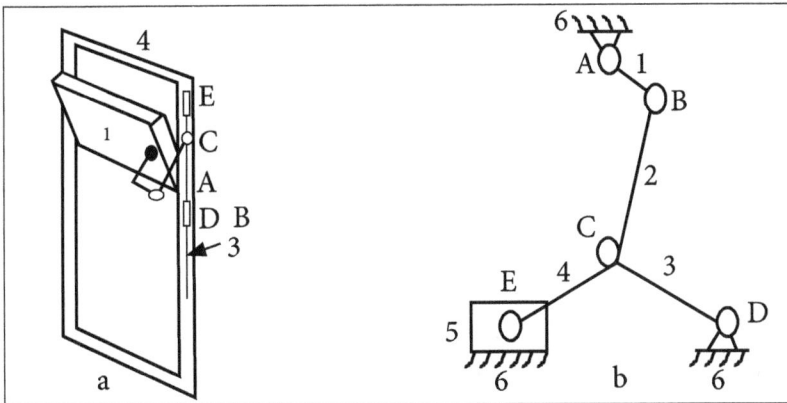

(b) Transom mechanism.

In order to control a mechanism, the number of independent input motions must equal the numbers of degrees of freedom of the mechanism. For example, the transom in above the figure (a) has a single degree of freedom, so it needs one independent input motion to open or close the window. E.g. push or pull rod 3 to operate the window.

Another example, the mechanism in above the figure (b) also has 1 degree of freedom. If an independent input is applied to link 1 (e.g., a motor is mounted on joint A to drive link 1), the mechanism will have a prescribed motion.

Grashof's law states that for a planar four bar linkage, the sum of the shortest and the longest link lengths cannot be greater than the sum of the remaining two link lengths if there has to be continuous relative motion between them.

Let us Consider a four-bar-linkage. Denote the smallest link by S , the longest link by L and the other two links by M_1 and M_2 . Given by,

$$L + S < M_1 + M_2$$

Then depending whether S is connected to the ground by one end, two ends or no end, the mechanism has the following type:

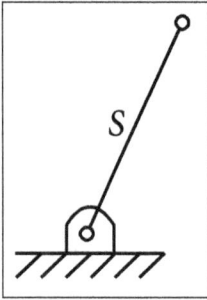   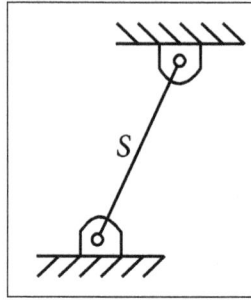   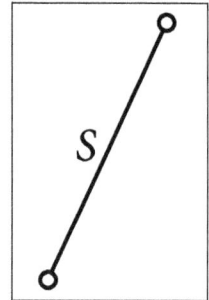

Crank – Rocker   Crank - Crank   Rocker – Rocker

$$L + S > M_1 + M_2,$$

Then the mechanism is of a rocker to rocker type.

Problem

M_1, M_2, M_3 and M_4 are four-bar linkages as shown in figure. The numbers on the figure indicate the respective link lengths in cm. Let us identify the nature of the mechanism, i.e., whether double crank, crank rocker or double rocker.

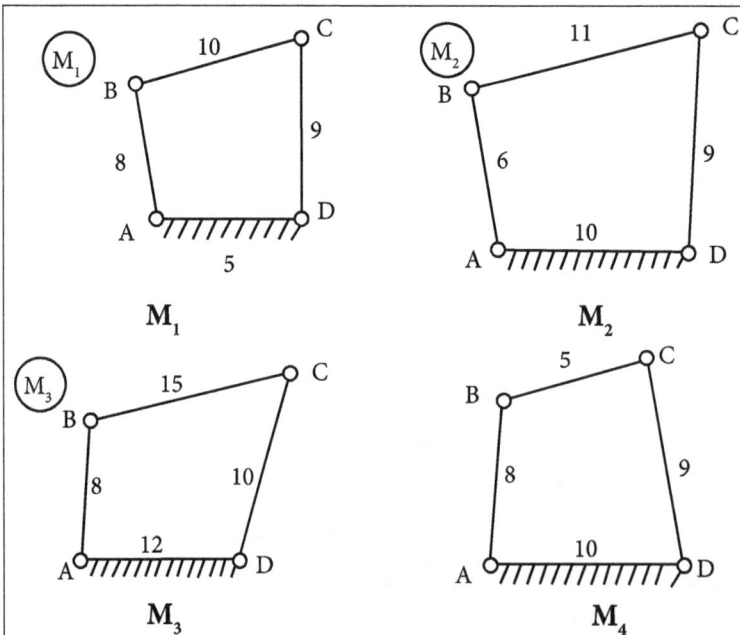

Solution:

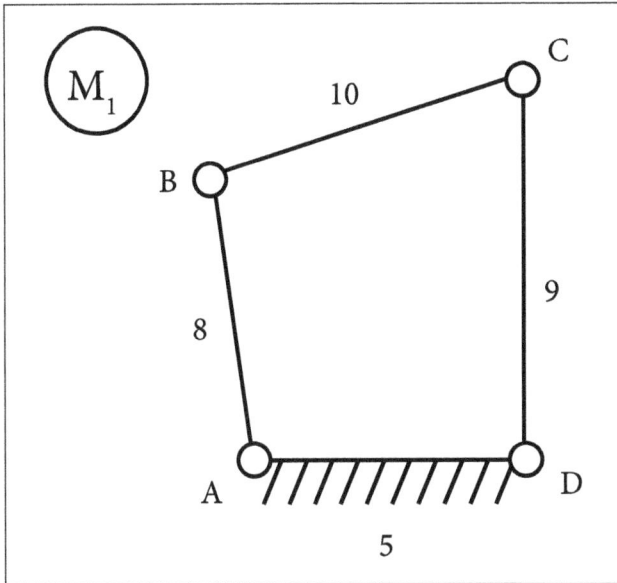

(i) AD = 5 cm,

AB = 8 cm,

BC = 10 cm,

CD = 9 cm.

Let,

l = Length of the longest link = 10 cm,

S = Length of the shortest link = 5 cm,

p, q = Length of the other two links = 8 cm and 9 cm,

We know that, if (l + s) (p + q), then the linkage is known as Grashoff's linkage.

Here, (10 + 5) (8 + 9), so given linkage is the Grashoff's linkage.

(ii) D = 10 cm,

AB = 6 cm,

BC = 11 cm,

CD = 9 cm.

Let, l = Length of the longest link = 11 cm,

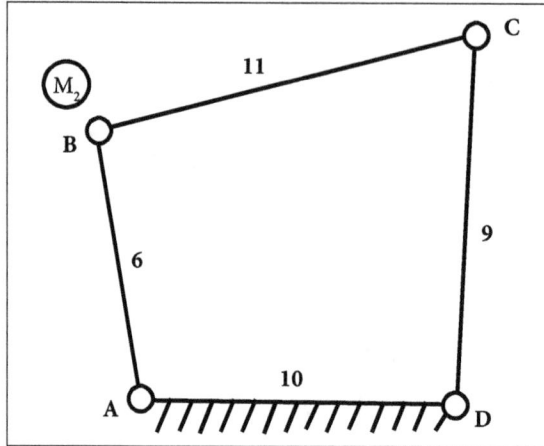

S = Length of the shortest link = 6,

p, q = Length of the other two link = 10 cm and 9 cm.

If $(l + s)$ $(p + q) = (11 + 6)$ $(10 + 9)$, so given linkage is the Grashoff's linkage.

(iii) AD = 12 cm,

AB = 8 cm,

BC = 15 cm,

CD = 10 cm.

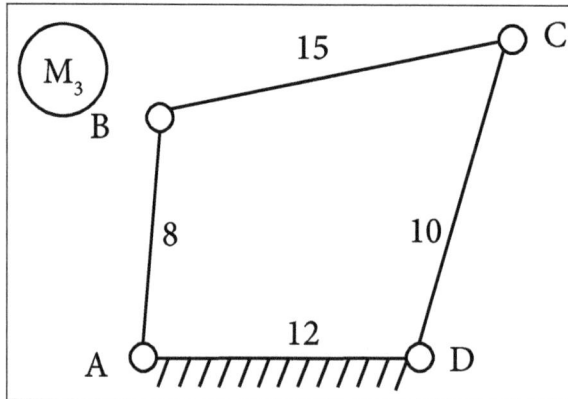

Let,

l = Length of the longest link = 15 cm,

S = Length of the shortest link = 8 cm,

p, q = Length of the other two link = 2 cm and 10 cm.

If $(l + s) < (p + q) = (15 + 8)$ and $(12 + 10)$.

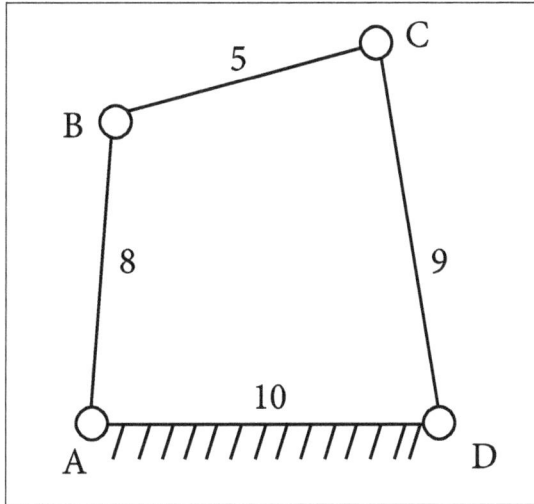

Let,

> l = Length of longest link = 10 cm,
>
> S = Length of the shortest link = 5 cm,
>
> p, q = Length of the other two link = 8 cm and 9 cm.
>
> If (l + s) < (p + q) = (10 + 5) & (8 + 9).

M_1 and M_2 are Grashoff's linkage. For Grashoff's linkage,

The following three mechanisms are possible:

- A double-crank mechanism when 'S' is the frame.
- Two different crank-rocker mechanism, when 'S' is crank and any one of adjacent link is frame.
- One double-rocker mechanism, when 'S' is the coupler (opposite to the frame).

1.2.4 Inversion and Machine

A mechanism can be defined as a combination of resistant bodies connected in such a way that they have a relative motion with each other.

A machine is a collection of several mechanisms which transmits force from power source to perform some useful work.

Kinematic Inversion

Kinematic inversion is defined as the process of fixing different links of a kinematic chain one at a time to produce different mechanisms.

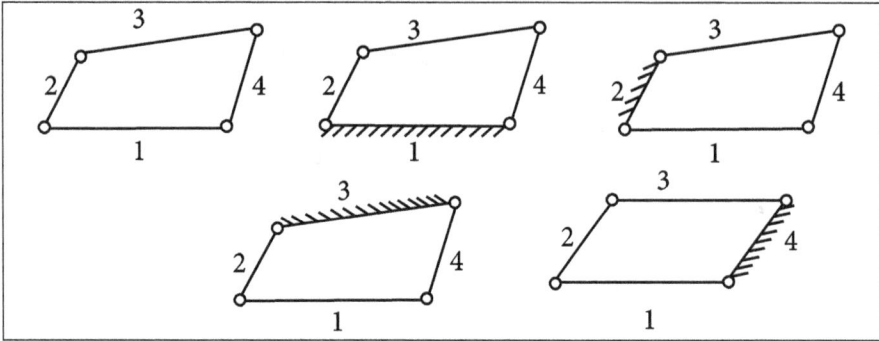

Inversions of 4R-Kinematic Chain

All the four inversions of 4R-kinematic chain are identical. By altering the proportions of lengths of links 1, 2, 3 and 4 respectively several mechanisms are obtained.

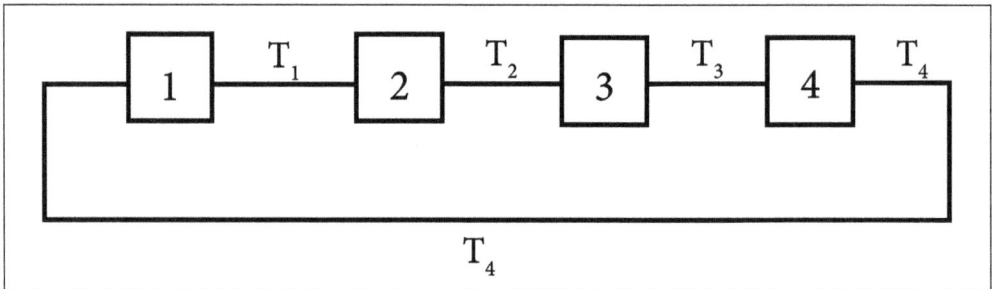

First Inversion

Crank-Rocker Mechanism

In this case for every complete rotation of link 2 (called a crank), the link 4 (called a lever or rocker), makes oscillation between extreme positions O_4B_1 and O_4B_2.

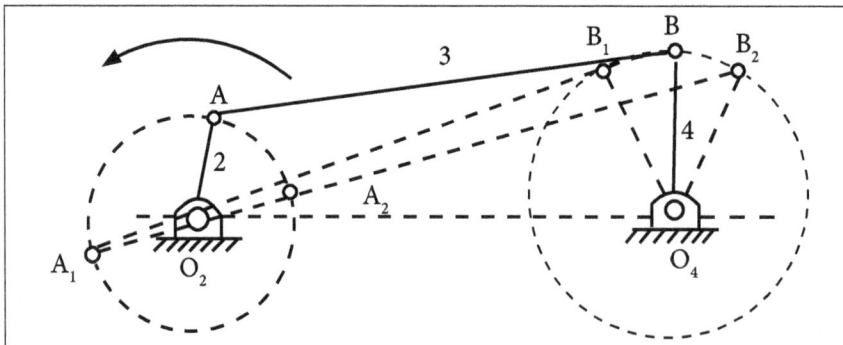

Crank-rocker Mechanism.

The position of O_4B_1 is obtained when point A is A_1 whereas position O_4B_2 is obtained when A is at A_2. It may be observed that crank angles for the two strokes (forward and backward) of oscillating link O_4B are not same. It may also be noted that the length of the crank is very short. If I_1, I_2, I_3 and I_4 are lengths of links 1, 2, 3 and 4 respectively, the proportions of the link may be as follows:

$$(l_1 + l_2) < (l_3 + l_4)$$
$$(l_2 + l_3) < (l_1 + l_4)$$

Double-lever Mechanism

In this mechanism, both the links 2 and 4 can only oscillate. Link O_2A oscillates between positions O_2A_1 and O_2A_2 whereas O_4B oscillates between positions O_4B_1 and O_4B_2. Position O_4B_2 is obtained when O_2A and AB are along straight line and position O_2A_1 is obtained when AB and O_4B are along straight line.

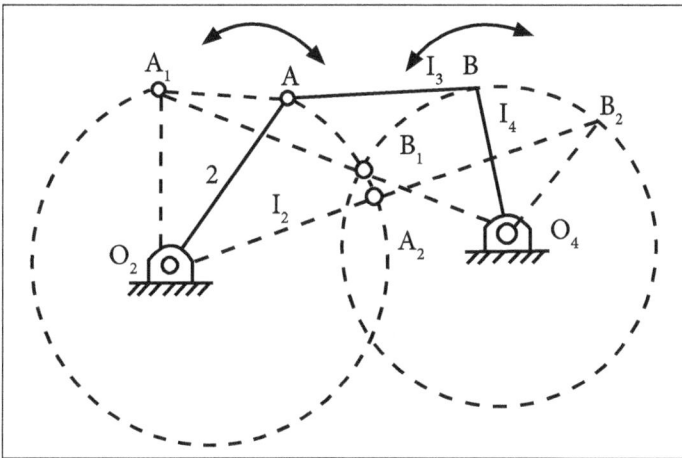

Double-lever Mechanism.

$$(l_3 + l_4) < (l_1 + l_2)$$
$$(l_2 + l_3) < (l_1 + l_4)$$

It may be observed that link AB has shorter length as compared to other links. If links 2 and 4 are of equal lengths and $I_1 > I_3$, this mechanism forms automobile steering gear.

Double Crank Mechanism

The links 2 and 4 of the double crank mechanism make complete revolutions.

There are two forms of this mechanism.

- Parallel Crank Mechanism

- Drag Link Mechanism

Parallel crank mechanism: In this mechanism, lengths of links 2 and 4 are equal. Lengths of links 1 and 3 are also equal.

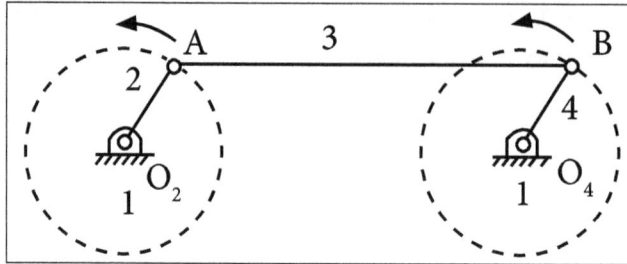

Double Crank Mechanism.

The familiar example is coupling of the locomotive wheels, where wheels act as cranks of equal length and length of the coupling rod is equal to centre distance between the two coupled wheels.

Drag link mechanism: In this mechanism also links 2 and 4 make full rotation. As the link 2 and 4 rotate sometimes link 4 rotate faster and sometimes it becomes slow in rotation.

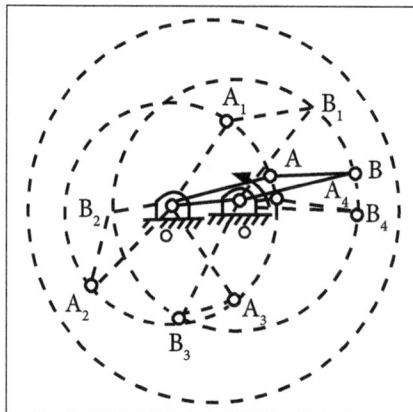

Drag Link Mechanism.

$$l_3 > l_1; l_4 > l_2$$

$$l_3 > (l_1 + l_4 - l_2)$$

$$l_3 < (l_2 + l_4 - l_1)$$

The length of link 1 is smaller as compared to other links.

1.3   Kinematic Chains and Inversions

A mechanism is one in which one of the links of a kinematic chain is fixed. Different mechanisms can be obtained by fixing different links of the same kinematic chain. These are called as inversions of the mechanism. By changing the fixed link, the number of mechanisms which can be obtained is equal to the number of links.

Excepting the original mechanism, all other mechanisms will be known as inversions of original mechanism. The inversion of a mechanism does not change the motion of its links relative to each other.

1.3.1 Inversions of Four Bar Chain and Single Slider Crank Chain

This mechanism is used to convert the rotary motion into reciprocating motion.

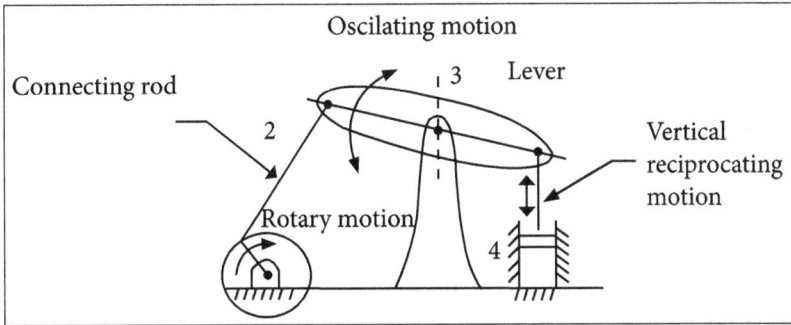

Inversion of four bar chain.

Inversions of Four Bar Chain

Important inversions of four bar chain are:

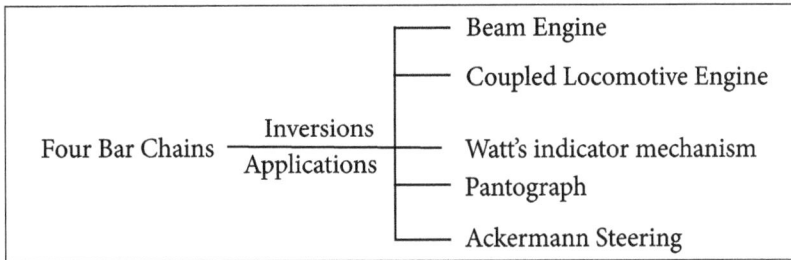

First Inversion: Crank and Lever Mechanism

As shown in figure (a), link 1 is the crank, link 4 is fixed and link 3 oscillates whereas in figure (b), link 2 is fixed and link 3 oscillates. The mechanism is also known as crank-rocker mechanism or a crank-lever mechanism or a rotary-oscillating converter.

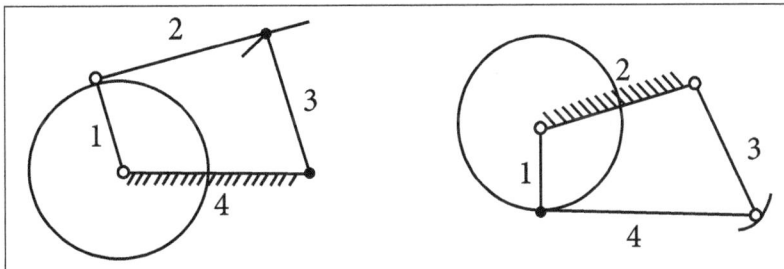

Crank and Lever Mechanism.

Example: Beam Engine

Application of beam engine: This is an example of crank-lever mechanism, where one link oscillates, while the other rotates about the fixed link, as shown in figure below,

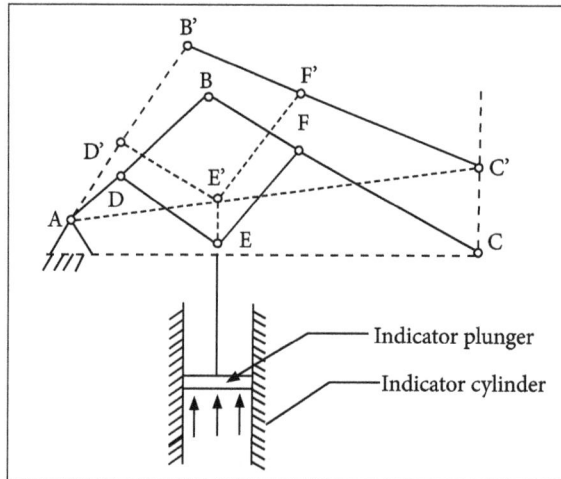

Beam Engine.

Watt's engine indicator (double lever mechanism) Panto graph:

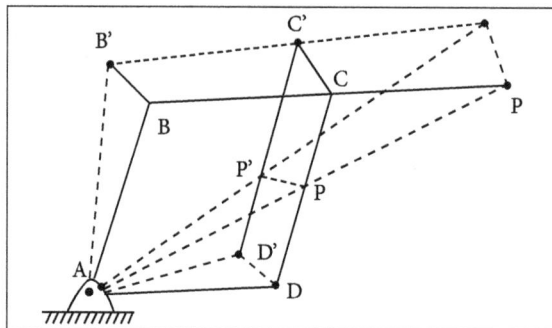

Pantograph.

Panto graph is a device which is used to reproduce a displacement exactly in an enlarged or reduced scale. It is used in drawing offices, for duplicating the drawings, maps, plans, etc. As shown in figure, it is a four bar mechanism in the form of a parallelogram ABCD with link BC extended to P.

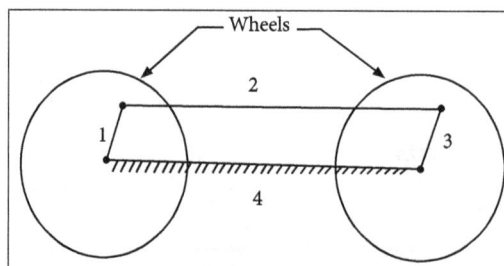

Mechanism of Pantograph.

This mechanism is used to transmit rotary motion from one wheel to the other wheels.

Second Inversion: Double Crank Mechanism

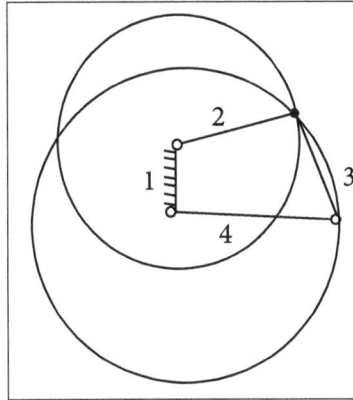

Double Crank Mechanism.

If the shortest link, i.e., link 1 (crank) is fixed, the adjacent links 2 and 4 would make complete revolutions, as shown in figure above. The mechanism thus obtained is known as crank-crank or double crank mechanism or rotary to rotary converter.

Coupling rod of a Locomotive: This is an example of a double crank mechanism where both cranks rotate about the points in the fixed link. It consists of four links. The opposite links are equal in length, since links 1 and 3 work as two cranks, the mechanism is also known as rotary-rotary converter.

Third Inversion

If the link opposite to shortest link is fixed, i.e., link 3 fixed, then the shortest link (link 1) is made coupler and the other two links 2 and 4 would oscillate. The mechanism thus obtained is known as rocker- rocker or double-rocker or double-lever mechanism or an oscillating-oscillating converter.

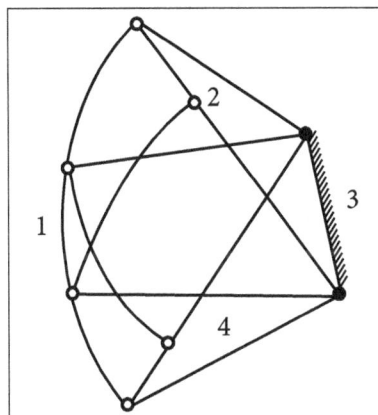

Third Inversion.

Sliding Connectors

Sliding connectors are used when one slider (the input) is to drive another slider (the output). Usually the two sliders operate in the same plane but in different directions.

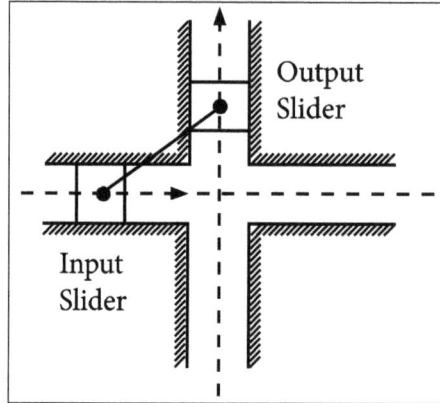

Sliding Connectors.

The figure shows a sliding connector which is obtained by a rigid link pivoted at each side of a slider. This is called as double slider crank mechanism.

Inversions of Slider Crank Chain

Kinematic Inversion

The process of fixing different links of a kinematic chain one at a time to produce distinct mechanisms is called kinematic inversion. Here the relative motions of the links of the mechanisms remain unchanged.

The Inversion of Slider Crank Chain: A slider-crank chain has the following inversions.

First Inversion

This inversion is obtained when link 1 is fixed and links 2 and 4 are made the crank and the slider respectively as shown in figure (1(a)).

Applications: Reciprocating engine, reciprocating compressor.

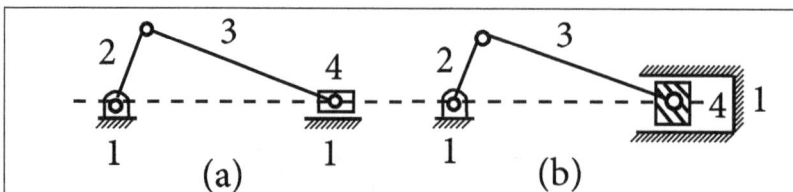

(1) First inversion.

As shown in figure (1(b)), if it is a reciprocating engine, 4 (piston) is the driver and if it is a compressor, 2 (Crank) is the driver.

Second Inversion

Firing of link 2 of a slider-crank chain results in the second inversion. The slider-crank mechanism of figure (1(a)) can also be drawn as shown in figure (2(a)). Further, when its link 2 is fixed instead of link 1, link 3 along with the slider at its end B becomes a crank. This makes link 1 to be rotated about o along with the slider which also reciprocates on it as shown in figure (2(b)).

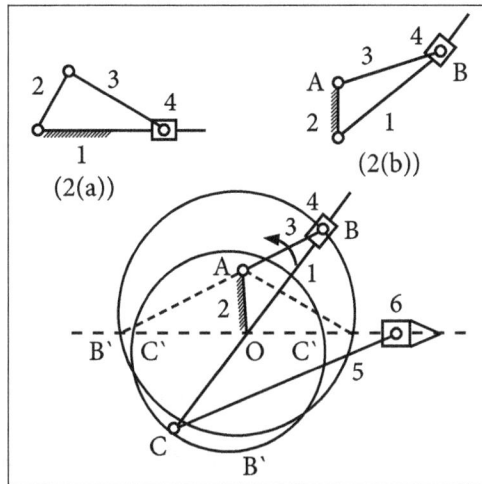

Whit Worth Quick - Return Mechanism.

Initially, let the slider 4 be at B' so that C be at C'. Cutting tool 6 will be in the extreme left position. With the movement of the crank, the slider traverses the path B'BB" whereas point C moves through C'CC". Cutting tool 6 will have the forward stroke. Finally, slider B assumes the position BB" and cutting tool 6 is in the extreme right position. The time taken for the forward stroke of slider 6 is proportional to the obtuse angle B"AB' at A.

Similarly, slider 4 completes the rest of the circle through path B"B'"B' and C passes through C"C'"C'. There is backward stroke of tool 6. The time taken is proportional to the acute angle B"AB' at A.

Let,

θ - Obtuse angle B'AB" at A,

β - Acute angle B'AB" at A.

Then , $\dfrac{\text{Time of Cutting}}{\text{Time of Re turn}} = \dfrac{\theta}{\beta}$

Third Inversion

By fixing link 3 of the slider crank mechanism, third inversion is obtained. Now, link 2 again acts as a crank and link 4 oscillates.

Applications

Oscillating cylinder engine, crank and slotted-lever mechanism.

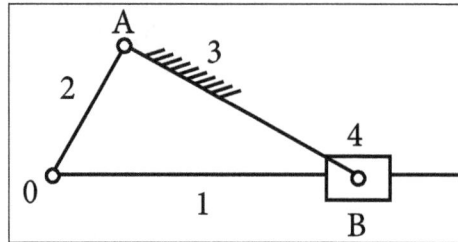

Crank mechanism.

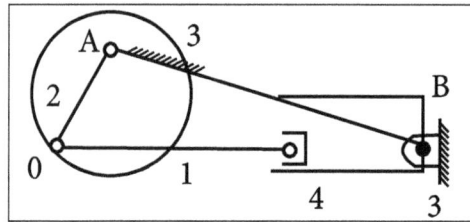

Slotted-lever mechanism.

Oscillating Cylinder Engine: As shown in figure above, link 4 is made in the form of a cylinder and a piston is fixed to the end of link 1.

The piston reciprocates inside the cylinder pivoted to the fixed link 3. The arrangement is known as oscillating cylinder engine, in which as the piston reciprocates in the oscillating cylinder, the crank rotates.

Fourth Inversion

If link 4 of the slider-crank mechanism is fixed, the fourth inversion is obtained. Link 3 can oscillate about the fixed pivot B on link 4. This makes end A of link 2 to oscillate about B and end O to reciprocate along the axis of the fixed link 4.

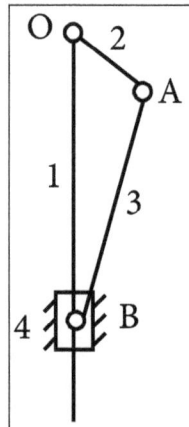

Fourth Inversion.

Application of Hand Pump

The below figure shows a hand pump. Link 4 is made in the form of a cylinder and a plunger fixed to the link 1 reciprocates in it.

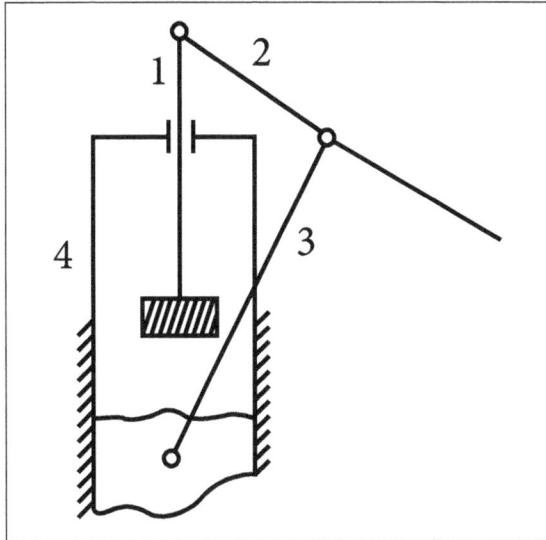

Hand pump.

Single Crank and Slotted Lever Mechanism

In this mechanism, the link AC (i.e., link 3) forming the turning pair is fixed, as shown in figure below.

The link 3 corresponds to the connecting rod of a reciprocating steam engine. The driving crank CB revolves with uniform angular speed about the fixed center C. A sliding block attached to the crank pin at B slides along the slotted bar AP and thus causes AP to oscillate about the pivoted point A.

A short link PR transmits the motion from AP to the ram which carries the tool and reciprocates along the line of stroke R_1R_2. The line of stroke of the ram (i.e., R_1R_2) is perpendicular to AC produced.

In the extreme positions, AP_1 and AP_2 are tangential to the circle and the cutting tool is at the end of the stroke. The forward or cutting stroke occurs when the crank rotates from the position CB_1 to CB_2 (or through an angle) in the clockwise direction. The return stroke occurs when the crank rotates from the position CB_2 to CB_1 (or through angle α) in the clockwise direction. Since the crank has uniform angular speed, therefore given by,

$$\frac{\text{Time of cutting stroke}}{\text{Time of return stroke}} = \frac{\beta}{\alpha} = \frac{\beta}{360° - \beta}$$

Or,

$$= \frac{360° - \alpha}{\alpha}$$

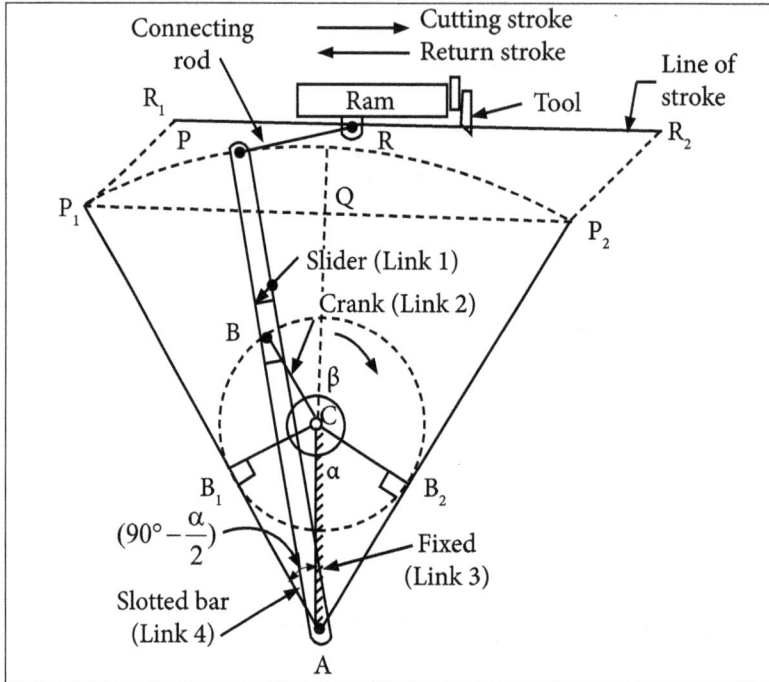

Crank and slotter lever quick return mechanism.

Since the tool travels a distance of $R_1 R_2$ during cutting and return stroke, therefore travel of the tool or length of stroke = $R_1 R_2 = P_1 P_2 = 2 P_1 Q = 2 AP_1 \sin P_1 AQ$.

$$= 2 AP_1 \sin\left(90° - \frac{\alpha}{2} \right) = 2 AP \cos \frac{\alpha}{2} \quad (\therefore AP_1 = AP)$$

$$= 2 AP \times \frac{CB_1}{AC} \dots \left(\because \cos\frac{\alpha}{2} = \frac{CB_1}{AC} \right)$$

$$= 2 AP \times \frac{CB}{AC} \dots \left(\because CB_1 = CB \right)$$

Therefore the return stroke is completed within shorter time. Thus it is called quick return motion mechanism.

1.3.2 Double Slider Crank Chain Mechanism

Obtaining different mechanisms by fixing different links in a kinematic chain is known as inversion of mechanism.

Elliptical Trammels

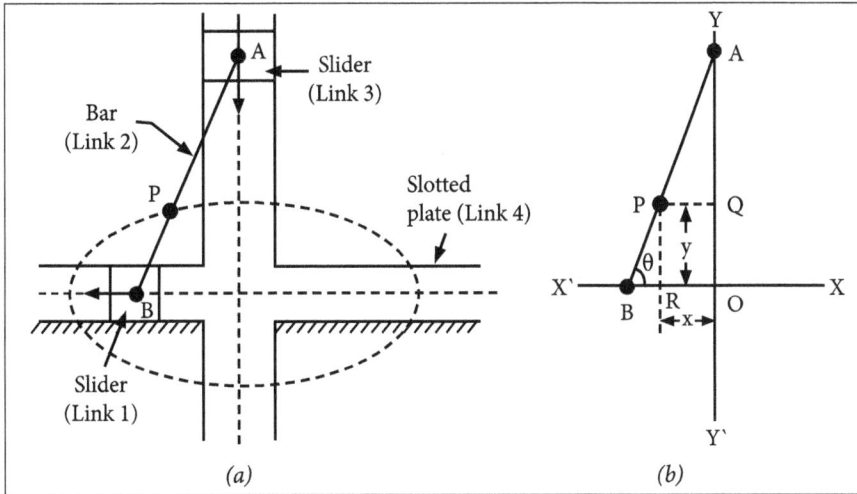

Elliptical Trammels.

Let us take OX and OY as horizontal and vertical axes and let the link BA is inclined at an angle θ with the horizontal, as shown in figure (b). Now the co-ordinates of the point P on the link BA will be,

$$x = AP\cos\theta; \text{ and } y = BP\sin\theta$$

Or,

$$\frac{x}{AP} = \cos\theta; \text{ and } \frac{y}{BP} = \sin\theta$$

Squaring and adding,

$$\frac{x^2}{(AP)^2} + \frac{y^2}{(BP)^2} = \cos^2\theta + \sin^2\theta = 1$$

This is the equation of an ellipse. Hence the path traced by point P is an ellipse whose semi-major axis is AP and semi-minor axis is BP.

Note: If P is the mid-point of link BA, then AP = BP.

The below equation can be written as,

$$\frac{x^2}{(AP)^2} + \frac{y^2}{(AP)^2} = 1 \text{ or } x^2 + y^2 = (AP)^2$$

This is the equation of a circle whose radius is AP. Hence if P is the midpoint of link BA, it will a circle.

Oldham's Coupling

The link 1 and link 3 form turning pairs with link 2. These flanges have diametrical slots cut in their inner faces, as shown in figure (b). The intermediate piece (link 4) which is a circular disc, have two tongues (i.e., diametrical projections) T_1 and T_2 on each face at right angle to each other, as shown in figure (c).

The tongues on the link 4 closely fit into the slots in the two flanges (link 1 and link 3). The link 4 can slide or reciprocate in the slots on the flanges.

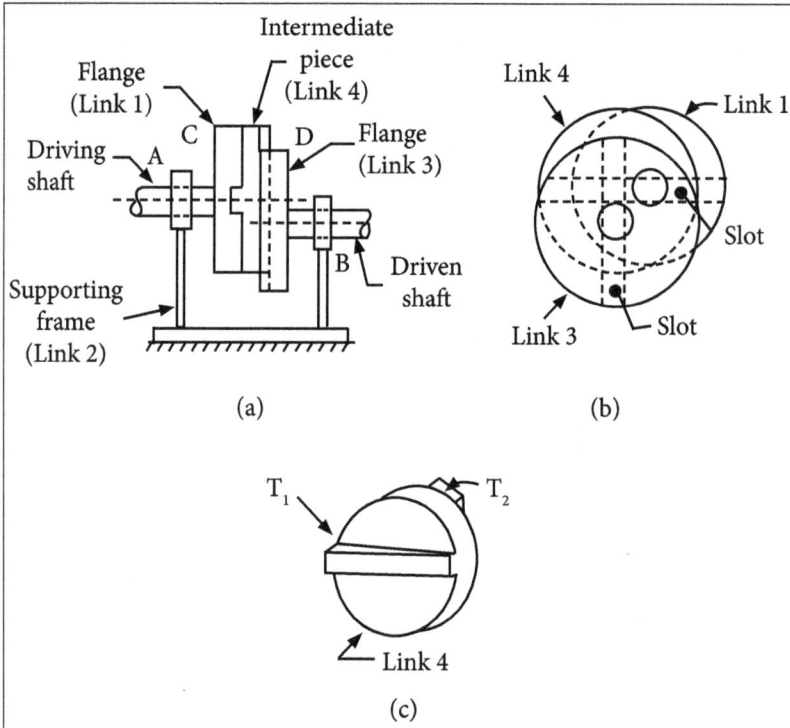

Oldham's Coupling

When the driving shaft A is rotated, the flange C (link 1) causes the intermediate piece (link 4) to rate at the same angle through which the flange has rotated and it further rotates the flange D (link 3) at the same angle and thus the shaft B rotates. Hence links 1, 3 and 4 have the same angular velocity at every instant. A little consideration will show that there is a sliding motion between the link 4 and each of the other links 1 and 3.

If the distance between the axis of the shafts is constant, the center of intermediate piece will describe a circle of radius equal to the distance between the axis of the two shafts. Therefore, the maximum sliding speed of each tongue along its slot is equal to the peripheral velocity of the center of the disc along its circular path.

Let,

ω = Angular velocity of each shaft in rad/s

r = Distance between the axis of the shaft in meters

i.e. Maximum sliding speed of each tongue (in m/s), $v = \omega r$

Scotch Yoke Mechanism

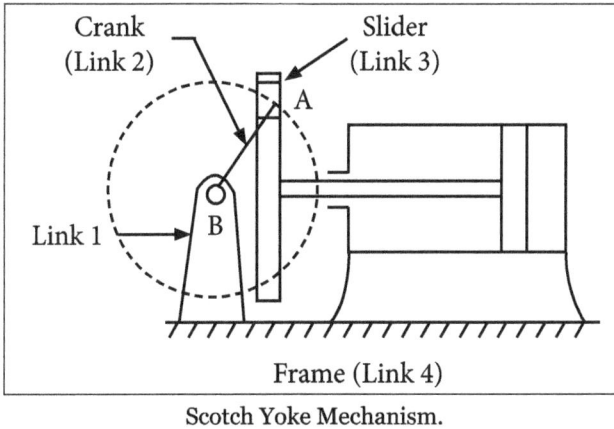

Scotch Yoke Mechanism.

This mechanism is used for converting rotary motion into a reciprocating motion. This inversion is obtained by fixing either the link 1 or link 3.

Link 1 $\Rightarrow$ fixed.

Link 2 $\Rightarrow$ crank.

Link 4 $\Rightarrow$ reciprocates.

Offset Slider Crank Mechanism

The offset slider crank mechanism is also used to get reciprocating motion of the slides. The figure Shows the outline sketch of an offset slider crank mechanism.

It has connecting rod (BC) hinged at point 'A'. The one end of a connecting rod is fitted with ram. The connecting rod can slide and turn about the point A, as shown in figure.

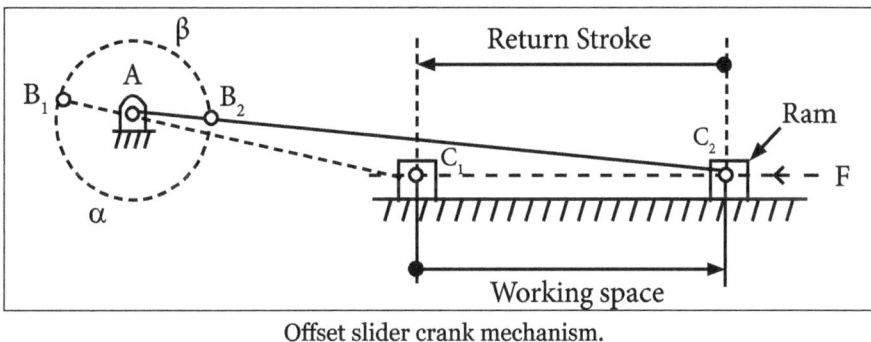

Offset slider crank mechanism.

When the connecting rod is at B, C, (shown in dotted line), then the ram is at the left extreme position C_1. As connecting rod slides and rotates through center point A, then the ram moves towards the right. When the connecting rod is at $B_2 C_2$, the ram is at the right extreme position C_2.

Thus the forward or cutting stroke occurs when the connecting rod rotates from the position AB_1 and AB_2 (or through an angle β) in the clockwise direction. The return stroke occurs when the crank rotates from the position AB_2 & AB_1 (or through angle α) in the clockwise direction.

$$\text{Then } \frac{\text{time of cutting stroke}}{\text{time of re turn stroke}} = \frac{\alpha}{\beta} = \frac{\alpha}{360-\alpha} \text{ or } = \frac{360-\beta}{\beta}$$

Problems

1. In a crank and slotted lever quick return motion mechanism, the distance between the fixed centers is 240 mm and the length of the crank is 120 mm. Let us determine the inclination of the slotted bar with the vertical in the extreme position and the time ratio of cutting stroke to the return stroke. If the length of the slotted bar is 450 mm let us also calculate the length of the stroke that passes through the extreme positions of the free end of the lever.

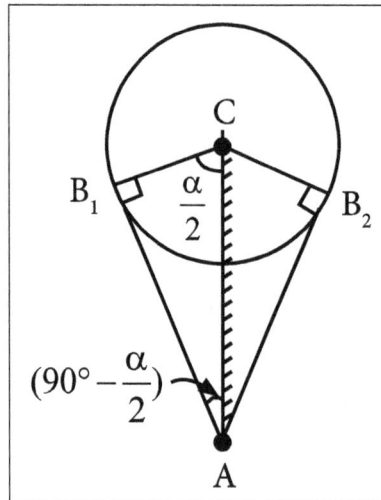

Solution:

Given:

AC = 240 mm,

CB_1 = 120 mm,

AP_1 = 450 mm,

Inclination of the slotted bar with the vertical.

Let, CAB_1 = Inclination of the slotted bar with the vertical.

$$\sin \lfloor CAB_1 = \sin\left(90^0 - \frac{\alpha}{2}\right)$$

The extreme positions of the crank are shown in figure. We know that,

$$= \frac{B_1 C}{AC} = \frac{120}{240} = 0.5$$

$$\lfloor CAB_1 = 90^0 - \frac{\alpha}{2}$$

$$= \sin^{-1} 0.5 = 30^0$$

Time ratio of cutting stroke to the return stroke:

We know that,

$$90^\circ = \alpha/2 = 30^\circ$$

$\therefore \alpha/2 = 90^\circ - 30^\circ = 60^\circ$ or $\alpha = 2 \times 60^\circ = 120^\circ$

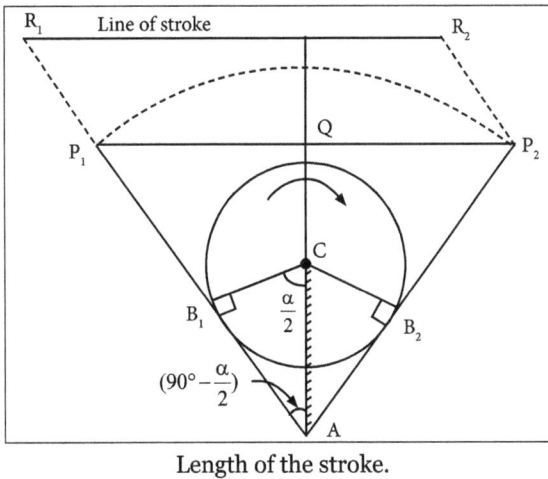

Length of the stroke.

We know that length of the stroke,

$$RlR_2 = P_1P_2 = 2P_1Q = 2AP_1 \sin (90^\circ - \alpha/2)$$

2. The length of the fixed link of a slotted lever mechanism is 250 mm and that of the crank 100 mm let us determine (i) The inclination of the slotter lever with the vertical in the extreme position. (ii) The ratio of the time cutting stroke to the time return of stroke. (iii) The length of the stroke, if the length of slotter lever is 450 mm and the line of stroke passing through the extreme position of the free end of the lever.

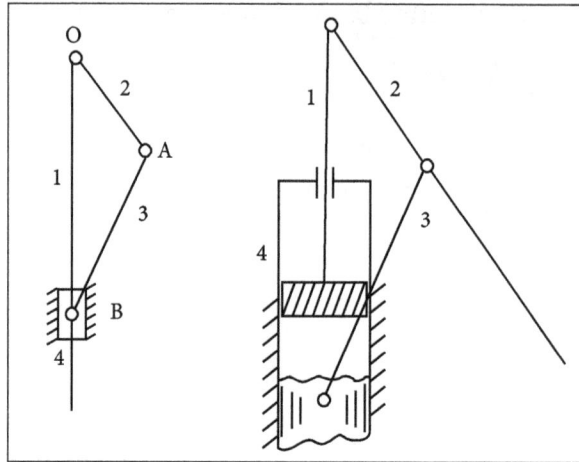

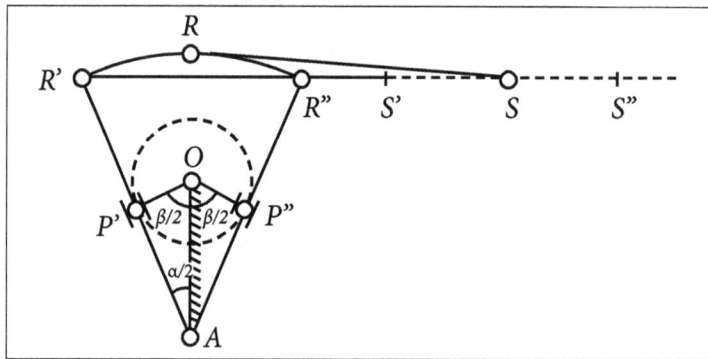

Solution:

Given:

OA =250mm OP' = OP" = 100 mm AR'= AR" = ar = 450 mm

$$\cos \frac{\beta}{2} = \frac{OP'}{OA} = \frac{100}{250} = 0.4$$

$$\frac{\beta}{2} = 66.4° \text{ or } \beta = 132.8°$$

- Angle of the slotted lever with the vertical $\alpha/2 = 90° - 66.4° = 23.6°$

- $$\frac{\text{Time of cutting stroke}}{\text{Time of return stroke}} = \frac{360° - \beta}{\beta} = \frac{360° - 132.8°}{132.8°} = 1.71$$

- Length of stroke = S'S" = R'R" = 2 AR' x sin ($\alpha/2$)

  = 2 x 450 sin23.6°

  = 360.3mm.

Mechanisms

2

2.1 Quick Return Motion Mechanisms

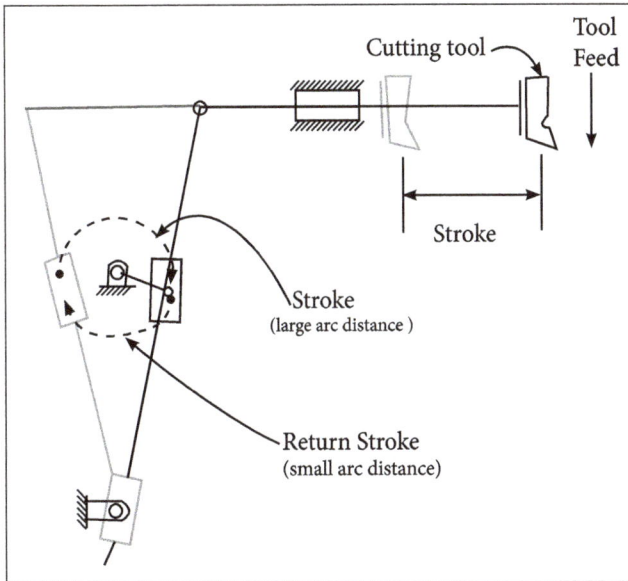

Quick return motion mechanisms.

Quick return mechanisms are used in machine tools such as shapers and power driven saws for the purpose of giving the reciprocating cutting tool a slow cutting stroke and a quick return stroke with a constant angular velocity of the driving crank. The ratio of time required for the cutting stroke to the time required for the return stroke is called the time ratio and is greater than unity.

2.1.1 Drag Link Mechanism

This is one of the inversions of four bar mechanism, with four turning pairs. Hence, link 2 is the input link, moving with constant angular velocity in anti-clockwise direction. Point C of the mechanism is connected to the tool post E of the machine. During cutting stroke, tool post moves from E_1 to E_2. The corresponding positions of C are C_1 and C_2 as shown in the figure below.

For the point C to move from C_1 to C_2, point B moves from B_1 to B_2, in anti-clockwise direction i.e., cutting stroke takes place when input link moves through angle $B_1 AB_2$ in anti-clockwise direction and return stroke takes place when input link moves through angle B_2AB_1 in anti-clockwise direction.

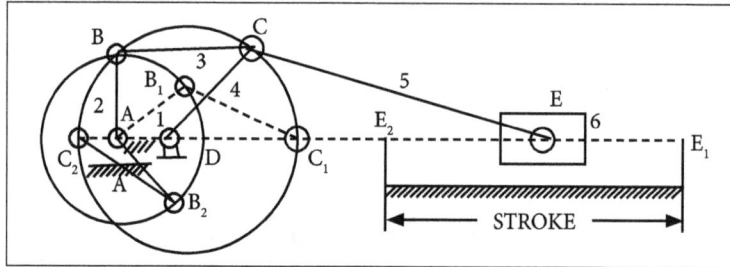

Drag Link Mechanism.

The time ratio is given by the following equation:

$$\frac{\text{Time for forward stroke}}{\text{Time for return stroke}} = \frac{B_1AB_2\left(\text{anti-clockwise}\right)}{B_2AB_1\left(\text{anti-clockwise}\right)}$$

2.1.2 Whitworth Mechanism

This is first inversion of slider mechanism, where, crank 1 is fixed. Input is given to link 2, which moves at constant speed. Point C of the mechanism is connected to the tool post D of the machine.

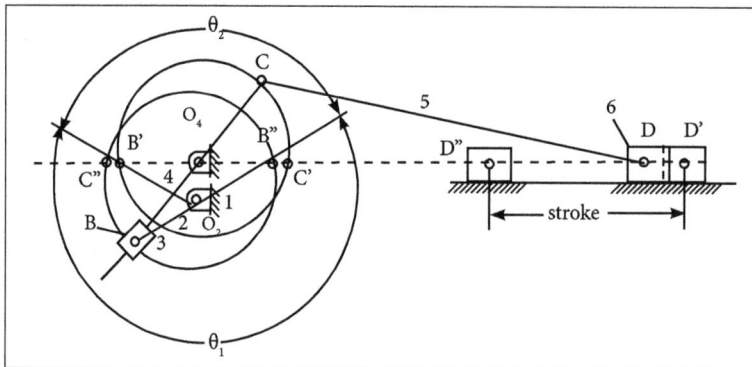

Whitworth mechanism.

During cutting stroke, tool post moves from D' to D". The corresponding positions of C are C' and C" as shown in the figure above. For the point C to move from C' to C", point B moves from B' to B", in anti-clockwise direction. I.E., cutting stroke takes place when input link moves through angle $B'O_2B"$ in anti-clockwise direction and return stroke takes place when input link moves through angle $B"O_2B'$ in anti-clockwise direction.

The time ratio is given by the following equation:

$$\frac{\text{Time for forward stroke}}{\text{Time for return stroke}} = \frac{B'O_2B"}{B"O_2B'} = \frac{\theta_1}{\theta_2}$$

Crank and Slotted Lever Quick Return Motion Mechanism

This is second inversion of slider mechanism, where, connecting rod is fixed. Input is given to link 2, which moves at constant speed.

Point C of the mechanism is connected to the tool post D of the machine. During cutting stroke, tool post moves from D' to D". The corresponding positions of C are C' and C" as shown in the fig. For the point C to move from C' to C", point B moves from B' to B", in anti-clockwise direction. i.e., cutting stroke takes place when input link moves through angle B'O$_2$B" in anti-clockwise direction and return stroke takes place when input link moves through angle B"O$_2$B' in anti-clockwise direction.

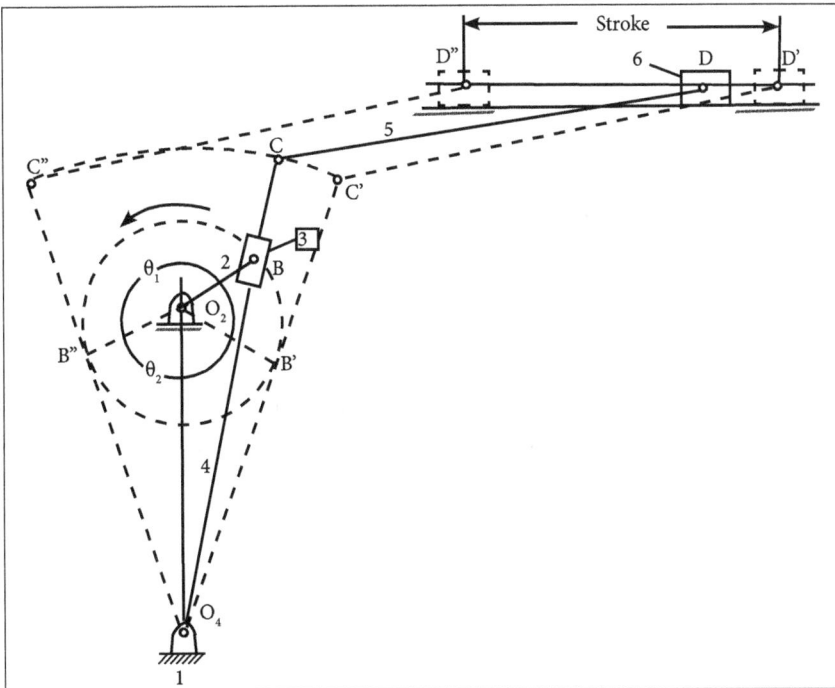

The time ratio is given by the following equation:

$$\frac{\text{Time for forward stroke}}{\text{Time for return stroke}} = \frac{B'O_2B''}{B''O_2B'} = \frac{\theta_1}{\theta_2}$$

2.2 Straight Line Motion Mechanisms

Sometimes a point on self recording instrument is required to move in a straight line. The obvious way of doing this is to use a sliding pair. But sliding pairs are bulky and gets rapidly worn out, so that in certain circumstances, it is desirable to obtain straight line motion by the use of turning pairs.

A mechanism which produces straight line motion by using turning pairs is known as straight line motion mechanism.

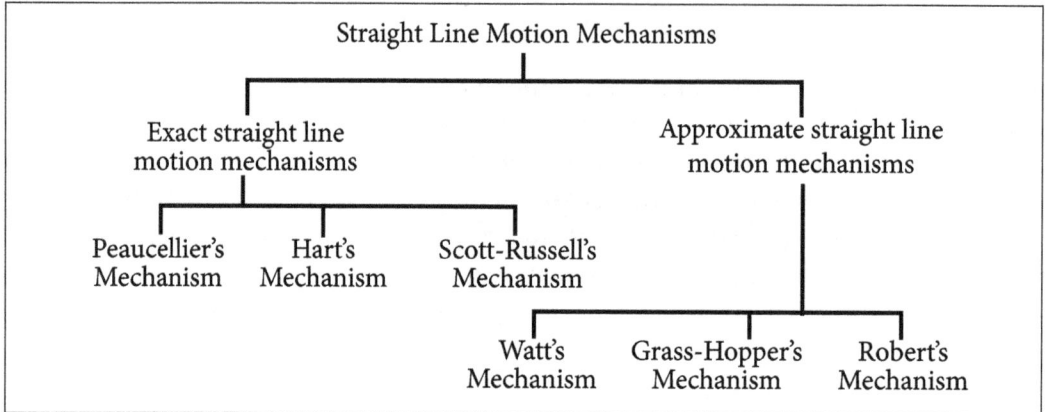

```
                        Straight Line Motion Mechanisms
                                      |
            ┌─────────────────────────┴─────────────────────────┐
     Exact straight line                              Approximate straight line
     motion mechanisms                                   motion mechanisms
            |                                                    |
   ┌────────┼────────────────┐                   ┌───────────────┼───────────────┐
Peaucellier's   Hart's   Scott-Russell's     Watt's      Grass-Hopper's     Robert's
 Mechanism    Mechanism   Mechanism        Mechanism       Mechanism        Mechanism
```

The straight line motion mechanisms can be further subdivided into two parts:

- Those in which straight line motion is exact or mathematically correct.
- Those in which the straight line motion produced is only approximate.

Condition for Exact Straight Line Motion

Let O, T and S be three distinct points of a mechanism, which lie in a straight line, for all configurations, as shown in the figure (a). The path of S will be a straight line perpendicular to the horizontal diameter OH of the circle, along the circumference of which the point T moves, if the product of OT and OS remains constant.

From the figure it is easy to see that triangles OTH and OSX are similar.

Hence, $\dfrac{OT}{OH} = \dfrac{OX}{OS}$

Since OT x OS is constant from the assumption made, it follows that the point S will move in a straight line SX.

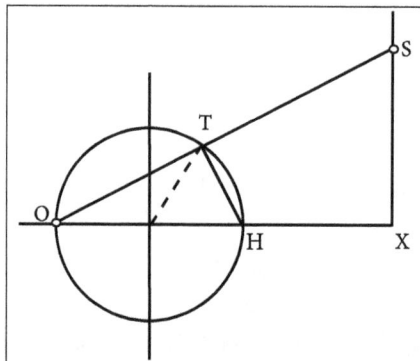

(a) Exact straight line motion.

Peaucellier and Hart's Mechanism

Peaucellier Mechanism

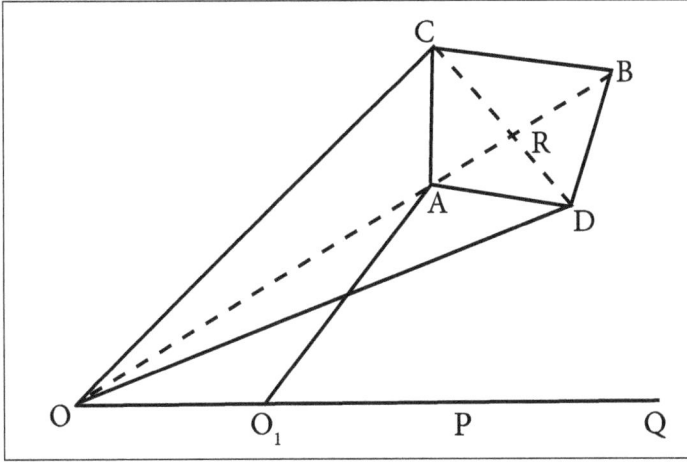

Peaucellier Mechanism.

It consists of a fixed link OQ and the other straight links O_1A, OC, OD, AD, DB, BC and CA are connected by turning pair at their intersections. The pin at A is constrained to move along the circumference of a circle with the fixed diameter OP, by means of the line O_1A.

AC = CB = BD = DA

OC = OD

$OO_1 = O_1A$

It may be proved that the product OA × OB remains constant, when the link O_1A rotates. Join CD to bisect AB at R. Now from right angled triangles ORC and BRC,

We have,

$OC^2 = OR_2 + RC^2$

$BC^2 = RB_2 + RC^2$

$OC^2 - BC^2 = OR^2 - RB^2$

$= (OR + RB) (OR - RB)$

$= OB \times OA$

$=$ Constant.

Since OC and BC are of constant length, the product OB × OA remains constant. Hence the point B traces a straight path perpendicular to the diameter OP.

Hart's Mechanism

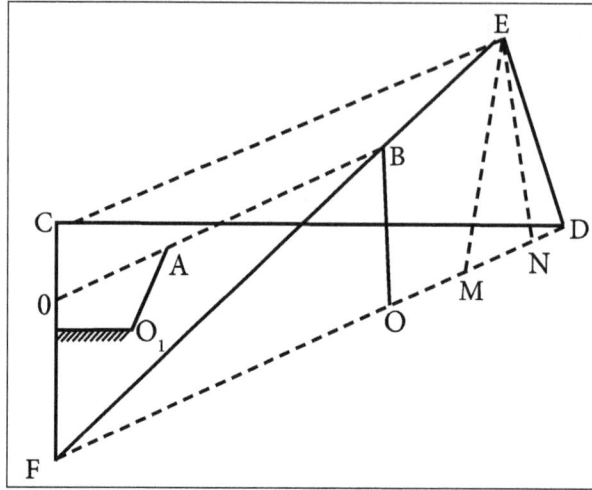

Hart's Mechanism.

This mechanism requires only six links as compared with the 8 links required by the Peaucellier mechanism.

It consists of fixed link OO_1 and other straight links O_1A, FC, CD, DE and EF are connected by turning pairs at their points for intersection FC = DE and CD = EF.

Hence OAB is a straight line. It may be proved now that the product OA × OB is constant.

From similar triangles CFE and OFB,

$$\frac{CE}{FC} = \frac{OB}{OF} \Rightarrow OB = \frac{CE \times OF}{FC}$$

From similar triangles FCD and OCA,

$$\frac{FO}{FC} = \frac{OA}{OC} \Rightarrow OA = \frac{FD \times OL}{FC}$$

$$OA \times OB = \frac{FD \times DC}{FC} \times \frac{CE \times OF}{FC}$$

= FD × CE × Constant.

[OC, OF and FC are fixed]

Now from point E, draw EM II^{el} to CF, EN is perpendicular to FD.

FD × CE = FD × FM

$$FD \times CE = FD \times FM$$

$$= (FN + ND)(FN - MN)\ [MN = ND]$$

$$= FN^2 - ND^2$$

$$= (FE^2 - NE^2) - (ED^2 - NE^2)$$

$$= FE^2 - ED^2$$

$$= \text{Constant}$$

$$\therefore OA \times OB = \text{Constant.}$$

If the Mechanism is pivoted about o as a fixed point and the point A is constrained to move on a circle with center o, then the point B will trace a straight line perpendicular to the diameter OP produced.

Problems

1. Let us design a pantograph for an indicator to be used to obtain the indicator diagram of an engine. The distance between the fixed point and the tracing point is 180 mm. The indicator diagram should be three times the gas pressure inside the cylinder of the engine.

Solution:

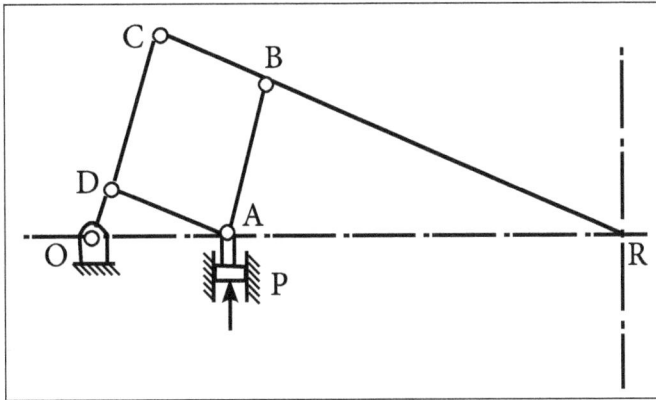

Refer the figure, OR = 180 mm and $\dfrac{OR}{OA} = 3$

Or, $\dfrac{180}{OA} = 3$ or OA = 60 mm

The relationship of the arms of a simplex indicator is as follows:

$$\frac{OR}{OA} = \frac{OC}{OD} = \frac{CR}{CB} = 3$$

Choose convenient dimensions of OD and DA. Let these be 30 mm and 50 mm respectively. Thus, as ABCD is to be a parallelogram and the above relation is to be fulfilled, the other dimensions will be,

$$OC = 30 \times 3 = 90 \text{ mm}; CR = 50 \times 3 = 150 \text{ mm}$$

Construct the diagram as follows:

- Locate D by making arc of radii 30 mm and 50 mm with centres O and A respectively.

- Produce OD to C such that OC = 90 mm.

- Join CR.

- Draw AB parallel to OC.

Thus, the required pantograph is obtained.

Crosby Indicator

This indicator employs a modified form of the pantograph. The mechanism has been shown in figure below:

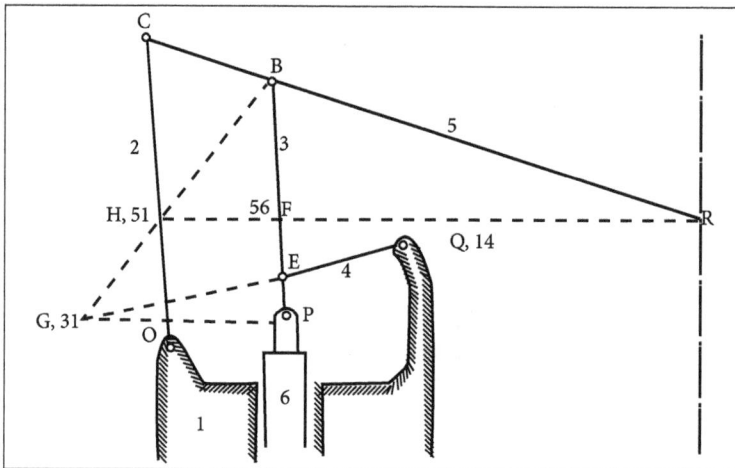

Crosby indicator.

To have a vertical straight line motion of R, it must remain in line with O and P, also the links OC and PB must remain approximately parallel.

As P lies on the link 3 and R on 5, locate the I-centres 31 and 51. If the directions of velocities of any two points on a link are known, the I-centre can be located easily which is the intersection of the perpendiculars to the directions of velocities at the two points.

First locate 31 as the directions of velocities of P and E on the link 3 are known.

The direction of velocity of P is vertical. Therefore 31 lie on a horizontal line through P.

The direction of velocity of E is perpendicular to QE. Therefore 31 lie on QE (or QE produced).

The intersection of QE produced with the horizontal line through P locates the point 31. Thus, the link 3 has its centre of rotation at 31 (link 1 is fixed) and the velocity of any point on the link is proportional to its distance from 31, the direction being perpendicular to a line joining the point with the I-centre.

To locate 51, the directions of velocities of B and C are known.

The direction of velocity of B is Perpendicular to 31 – B. Therefore, 51 lie on 31 – B.

The direction of velocity of C is Perpendicular to OC. Therefore, 51 lie on OC.

Thus, 51 can be located. Now, the link 5 has its centre of rotation at 51. The direction of velocity of point R on this link will be perpendicular to 51-R. To have a vertical motion of R, it must lie on a horizontal line through 51.

The ratio of the velocities of R and P is given by,

$$\frac{v_r}{v_p} = \frac{v_r}{v_b} \frac{v_b}{v_p} \quad \text{(B is common to 3 and 5)}$$

$$= \frac{51-R}{51-B} \cdot \frac{31-B}{31-P}$$

$$= \frac{51-R}{51-B} \cdot \frac{51-B}{51-F} \quad \left(\because \Delta s \text{ BPG and BFH are similar} \right)$$

$$= \frac{51-R}{51-F}$$

$$= \frac{CR}{CB} \quad \left(\because \Delta s \text{ CRH and BRF are similar} \right)$$

$$= \text{Constant}$$

This shows that the velocity or the displacement of R will be proportional to that of P.

Alternatively, locate the I-centre 56 by using Kennedy's theorem. It will be at the point F (the intersection of lines joining I-centres 16, 15 and 35, 36, not shown in the figure). First consider this point 56 to lie on the link 6. Its absolute velocity is the velocity of 6 in the vertical direction (1 being fixed). Now, consider the point 56 to lie on the link 5.

The motion of 5 is that of rotation about 51 (1 being fixed). Thus, velocity of R on the link 5 can be found as the velocity of 56, another point on the same link is known.

$$\frac{v_r}{v_f} = \frac{51-R}{51-F} = \frac{51-R}{51-F} \quad \left(v_f - v_p\right)$$

$$= \frac{CR}{CB}$$

Thomson Indicator

A Thomson indicator employs a Grass-Hopper mechanism OCEQ. R is the tracing point which lies on CE produced as shown in figure below.

The best position of the tracing point R is obtained as discussed.

Locate the I-centres 31 and 51 as in case of a Crosby indicator. The directions of velocities of two points C and E on the link 5 are known. Therefore, first locate the I-centre 51.

The direction of velocity of C is 1 to OC. Therefore 51 lie on OC.

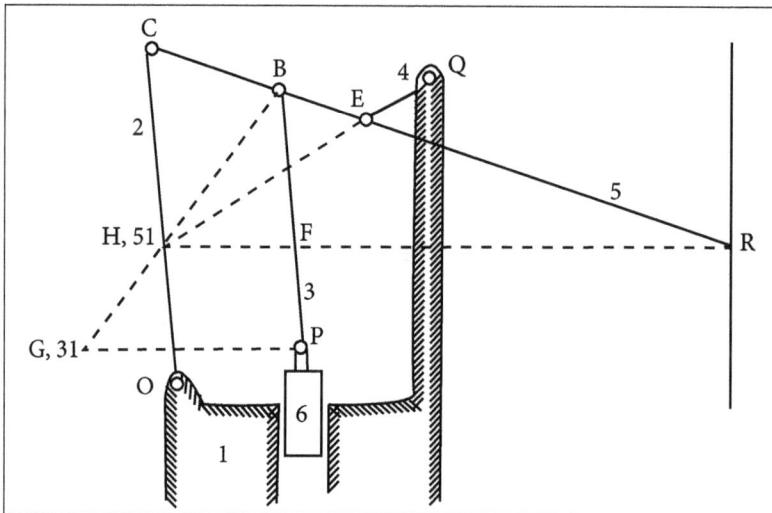

Thomson indicator.

The direction of velocity of E is 1 to QE, Therefore 51 lies on QE (or QE produced).

Thus, 51 can be located.

Now, the directions of velocities of two points B and P on the link 3 are known.

The direction of velocity of B is perpendicular to 51-B, (B is on the link 5 also) Therefore, 31 lies on the line 51-B. The direction of velocity of P is vertical. Therefore, 31 also lie on a horizontal line through P. Thus, 31 can be located.

As R is to move in a vertical direction, it must lie on a horizontal line through the

I-centre of the link 5 on which the pointer lies. Similar to the case of a Crosby indicator, the velocity ratio is given by,

$$\frac{v_r}{v_p} = \frac{CR}{CB} = \text{Constant}$$

Therefore, the velocity or the displacement of R is proportional to that of P. It is to be remembered that since OC and PB does not remain parallel for all positions, R moves in an approximate vertical line. However, the variations are negligible. Alternatively, the I-centre 56 can be located by using Kennedy's theorem. It will be at the point F.

Dobbie McInnes Indicator

This indicator is similar to a Thomson indicator, the difference being that the link 3 is pivoted to a point in link 4 instead of a point on link 5. Thus, the motion of the indicator piston is imparted to the link 4.

The indicator is shown in figure below. Locate the I-centre 51 as before. Locate R by finding the intersection of CE and a horizontal line through 51. Locate I-centre 31 as usual.

Now,

$$\frac{v_r}{v_p} = \frac{v_r}{v_e} \times \frac{v_e}{v_b} \times \frac{v_b}{v_p}$$

$$= \frac{51-R}{51-E} \cdot \frac{QE}{QB} \cdot \frac{31-B}{31-P}$$

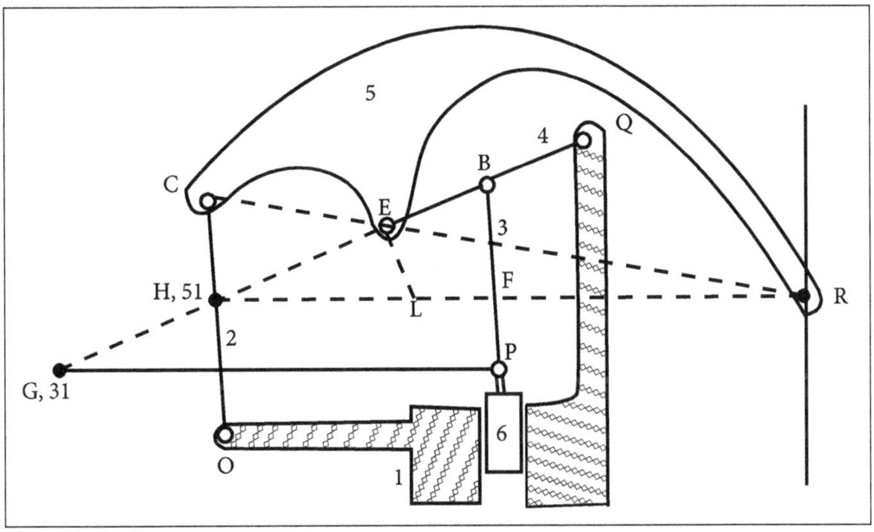

Dobbie McInnes indicator.

$$= \frac{51-R}{51-E} \cdot \frac{QE}{QB} \cdot \frac{51-B}{51-F} \quad (\because \Delta s \text{ BPG and BFH are similar})$$

$$= \frac{51-R}{51-E} \cdot \frac{QE}{QB} \cdot \frac{51-E}{51-L} \quad (\because \Delta s \text{ BFH and ELH are similar})$$

$$= \frac{51-R}{51-L} \cdot \frac{QE}{QB} = \frac{CR}{CE} \cdot \frac{QE}{QB} = \text{constant}$$

This expression also gives approximately the ratio of the displacement of R to that of P.

Problem

1. Peaucellier mechanism: Figure (a) shows the link MAC which oscillates on a fixed centre A. Another link OQ oscillates on centre O. The links AB and AC are equal. Also BP = PC = CQ = QB.

Let us Locate the position of O, when:

- P moves in a straight line.

- P moves in a circle with centre A.

- P moves in a circle with centre at OA produced.

- P moves in a circle with centre O and Q moves in a straight line (Modify the lengths of links if necessary.)

Solution:

- As in the Peaucellier mechanism, O is located by drawing a straight line through A and perpendicular to the motion of P such that AO = OQ [Figure (a)].

- If O is made to coincide with A, AQ would be equal to OQ. Thus, Q and P will be fixed on AP. Q will rotate about A and thus P will also rotate in a circle about A with AP as the radius [Figure (b)].

- From the above two cases, it can be observed that in (i) P moves in a circle with the centre at infinity on OA produced and in (ii) P moves in circle with the centre at A. Thus if P is to move in a circle with the centre in-between A and infinity on OA produced, O must lie in-between O and A or in other words OQ should be greater than OA [Figure(c)].

- The mechanism will be similar to the Peaucellier mechanism. P is to be joined with O by a link so that P moves in a circle about O and OA = OP. The lengths can be modified in two ways [Figure (d)].

  ○ OA is increased and OA and OP are made equal.

  ○ Lengths AB and AC are reduced in such a way that OA = OP.

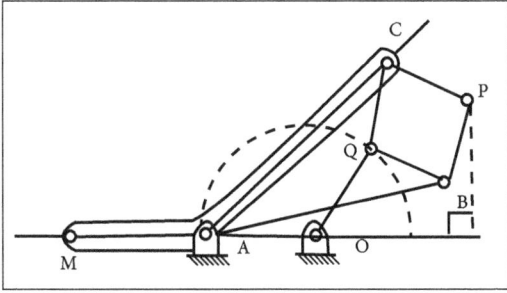

(a)

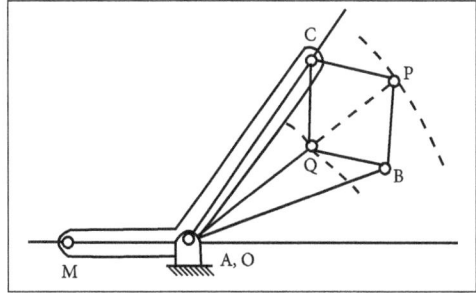

b)

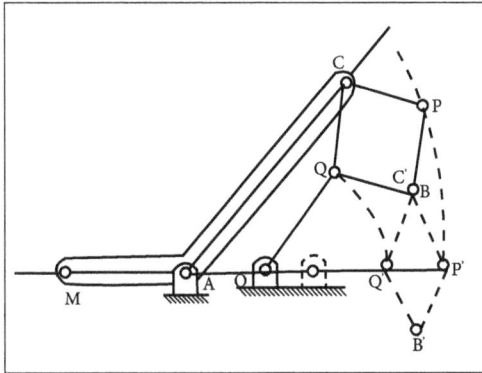

(c)

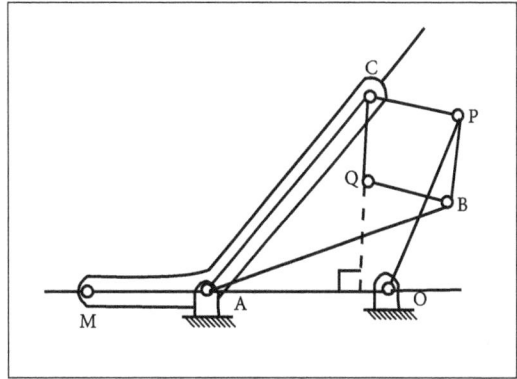

(d)

2. Mechanism: A circle with EQ' as diameter has a point Q on its circumference. P is a point on EQ produced such that if Q turns about E, EQ. EP is constant. Let us prove that the point P moves in a straight line perpendicular to EQ'.

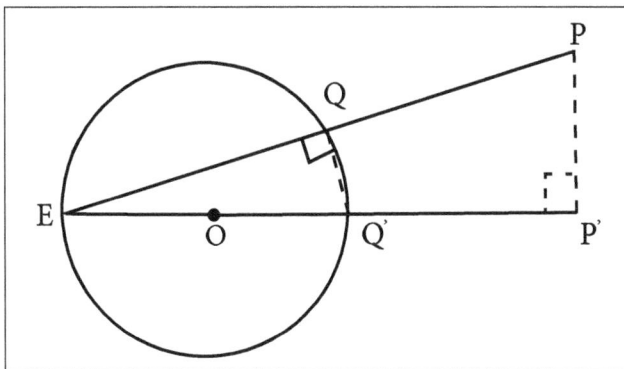

Solution:

Let PP' be perpendicular to EQ' produced (Figure).

For any position of Q on the circumference of the circle with diameter EQ', triangles EQQ' and EP'P are similar (∟QEQ' is common and ∟EQQ' = ∟EP'P = 90°).

$$\therefore \frac{EQ}{EQ'} = \frac{EP'}{EP} \text{ Or, } EQ'. EP' = EQ.EP$$

$$\text{Or, } EP' = \frac{EQ \cdot EP}{EQ'} = \text{cons} \tan t$$

EQ' is fixed and EQ. EP = Constant.

Thus, EP will be constant for all positions of Q. Therefore, the location of P' is fixed which means that P moves in a straight line perpendicular to EQ'.

2.2.1 Robert's Straight Line Mechanism

Like the Chebyshev's mechanism Robert's approximate straight line mechanism is a symmetrical four bar linkage.

The construction of Robert's mechanism is different from the approximate straight line mechanisms discussed so far, in the sense that, this mechanism has an extension to the coupler at the coupler mid-point. This extension is perpendicular to the line joining the two adjacent joints. The end point of the coupler extension generates an approximate straight line for the motion between the fixed pivots.

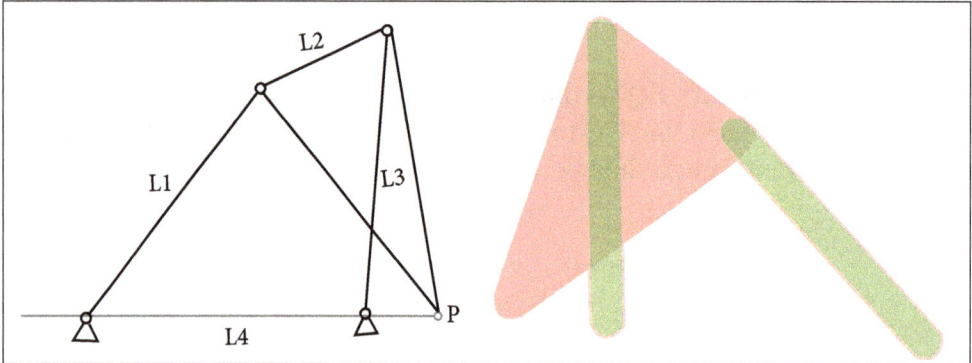

This approximate straight line mechanism is generally used for linear guidance of the tracing point. The point required to traverse on straight line is constrained to the end point of the coupler extension. Robert's straight line mechanism is normally used in the coupler driven mode, that is, the mechanism is not driven by either of the cranks or rockers instead the coupler extension is used to just guide the requisite point along an approximate straight line.

In Robert straight line mechanism, which is a four-bar linkage shown in the figure (a), the lengths of links 2 and 4 are equal. A point P on the perpendicular bisector of connecting rod 3 is to be chosen, such that it describes an approximate horizontal straight line.

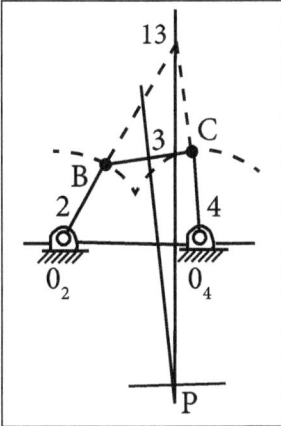

(a) Robert Straight Line Mechanism   (b) A Special Case of Robert's Straight line mechanism.

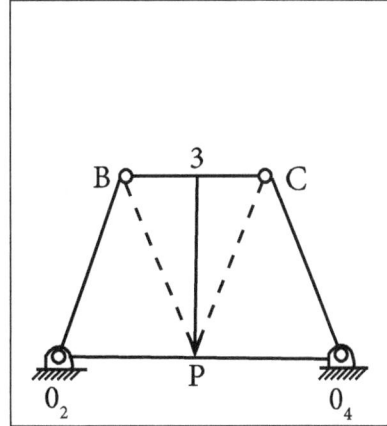

This can be best done by using instantaneous centre approach, as applied to Watt mechanism. The figure (a), illustrates this. A special case of Robert's straight line motion is shown in the figure (b), in which 2. It is evident that point P describes a straight line O_2PO_4.

$$BC = O_2O_4$$

2.3 Intermittent Motion Mechanisms

An intermittent-motion mechanism is a linkage which converts continuous motion into intermittent motion. These mechanisms are commonly used for indexing in machine tools.

Intermittent motion is a sequence of motions and dwells. A dwell is a period in which the output link remains stationary while the input link continues to move. There are many applications in machinery that require intermittent motion. The cam-follower variation on the four-bar linkage as shown in the figure below, is often used in these situations.

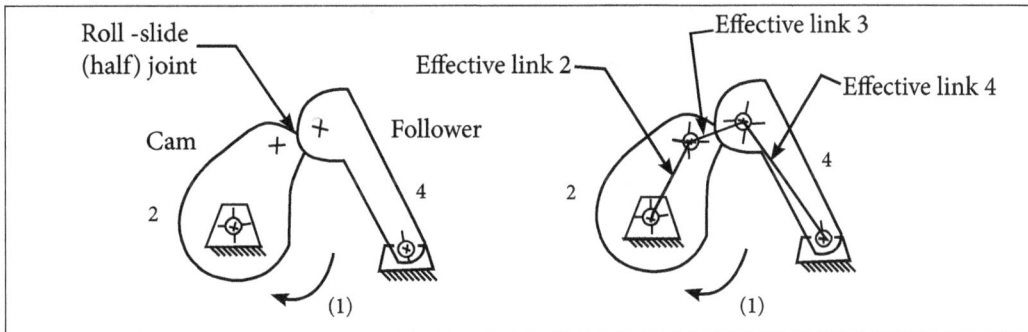

The cam-follower variation on the four-bar linkage.

2.3.1 Geneva Wheel Mechanism

The Geneva drive is also commonly called a Maltese cross mechanism. The Geneva mechanism translates a continuous rotation into an intermittent rotary motion.

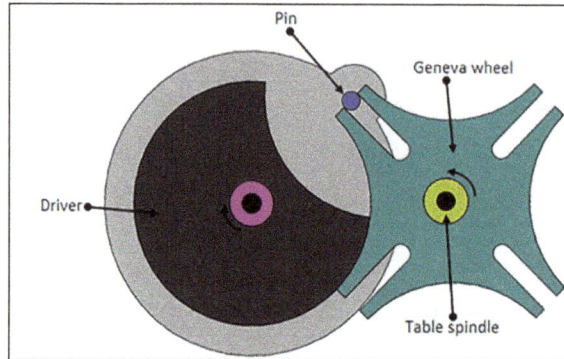

(a) Geneva wheel mechanism.

The rotating drive wheel has a pin that reaches into a slot of the driven wheel. The drive wheel also has a raised circular blocking disc that locks the driven wheel in position between steps as shown in the figure (a).

There are three basic types of Geneva motion mechanisms namely external, internal and spherical. The spherical Geneva mechanism is very rarely used. In the simplest form, the driven wheel has four slots and hence for each rotation of the drive wheel it advances by one step of 90°. If the driven wheel has n slots, it advances by 360°/n per full rotation of the drive wheel.

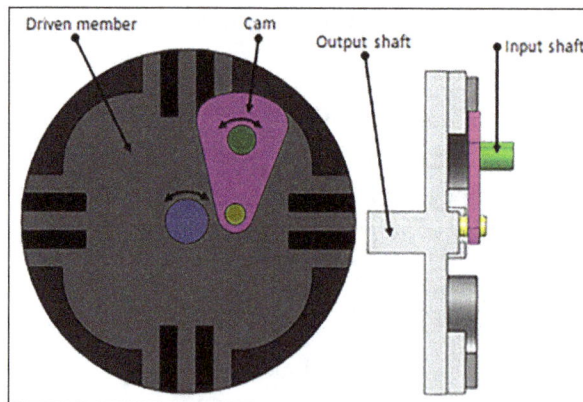

(b) Internal Geneva mechanism.

In an internal Geneva drive the axis of the drive wheel of the internal drive is supported on only one side as shown in the figure (b). The angle by which the drive wheel has to rotate to effect one step rotation of the driven wheel is always smaller than 180° in an external Geneva drive and is always greater than 180° in an internal one. The external form is the more common, as it can be built smaller and can withstand higher mechanical stresses.

Because the driven wheel always under full control of the driver, impact is a problem. It can be reduced by designing the pin in such a way that the pin picks up the driven member as slowly as possible. Both the Geneva mechanisms can be used for light and heavy duty applications. Generally, they are used in assembly machines.

Intermittent linear motion from rotary motion can also be obtained using Geneva mechanism as shown in the figure (c). This type of movement is basically required in packaging, stamping, assembly operations, and embossing operations in manufacturing automation.

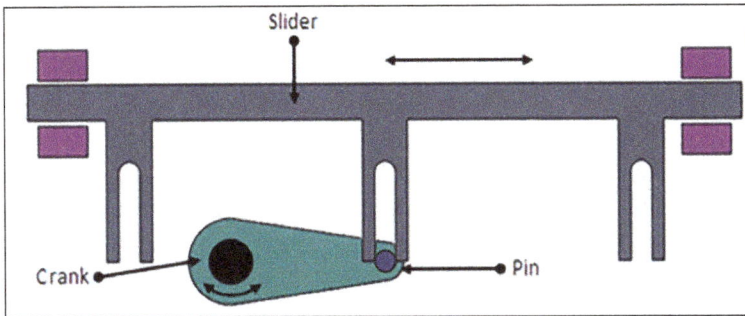

(c) Linear intermittent motion using Geneva mechanism.

2.3.2 Ratchet and Pawl Mechanism

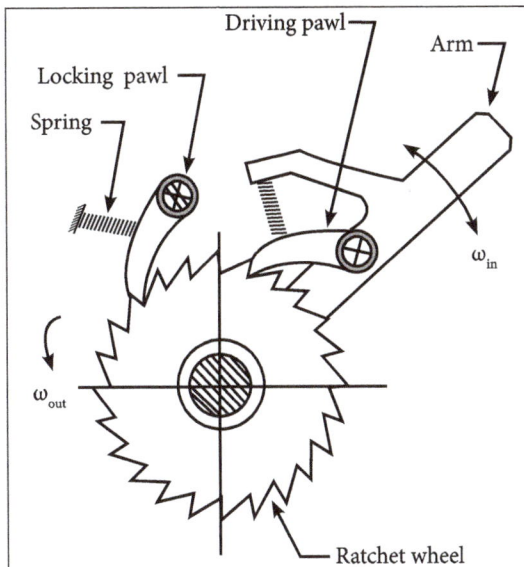

Ratchet and Pawl mechanism.

A ratchet is a device that allows linear or rotary motion in only one direction. The figure above shows a schematic of the same. It is used in rotary machines to index air operated indexing tables. Ratchets consist of a gearwheel and a pivoting spring loaded pawl that engages the teeth. The teeth or the pawl, are at an angle so that when the teeth are moving in one direction the pawl slides in between the teeth.

The spring forces the pawl back into the depression between the next teeth. The ratchet and pawl are not mechanically interlocked hence easy to set up. The table may over travel if the table is heavy when they are disengaged. Maintenance of this system is easy.

2.4   Toggle Mechanisms

The mechanical advantage of the 4-bar mechanism is directly proportional to the sine of the angle γ between the coupler and the follower link and inversely proportional to the angle β between the coupler and the driving link. As these angles go on changing, mechanical advantage also goes on changing with configuration of linkage.

Clearly, mechanical advantage becomes infinite when the sine of the angle between input link and coupler becomes zero. For such a position of driver and coupler (making β = 0 or 180°), only a small input torque is sufficient to overcome a large output load. In either of these positions, the mechanical advantage is infinite and a toggle position is said to be achieved for linkage.

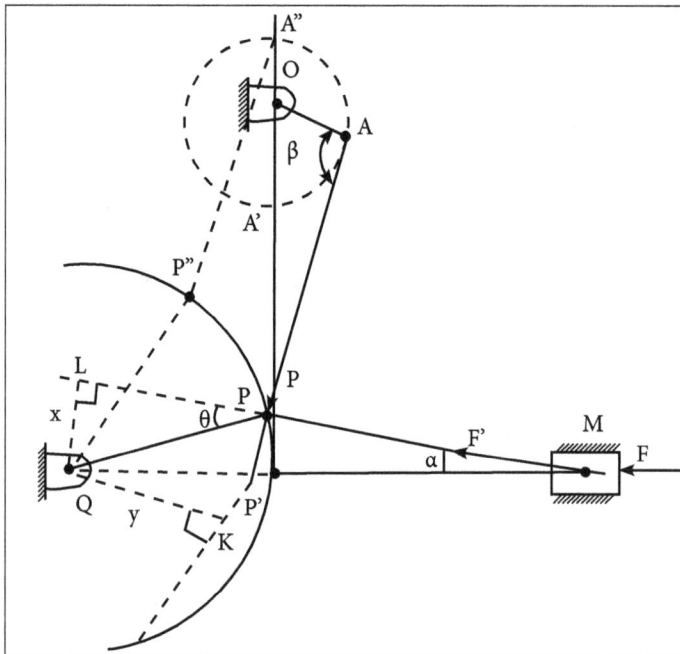

Toggle mechanism.

Note that these positions also correspond to extreme positions of rocker, which is the driven link. The figure shows one such toggle mechanism. As the slider M approaches its end of stroke to the right its velocity approaches zero. This is the position when β—> 180° and mechanical advantage approaches to infinity and the slider is capable of exerting very large force F when the applied force P is small. Taking moments about Q of force F' in connecting rod and tangential force P on crank,

$$F'(x) = P(y) \quad ....(i)$$

But,

$$F' = F/\cos\alpha \quad ....(ii)$$

Substituting in equation (ii)

Or,

$$\frac{F}{P} = \left(\frac{y}{x}\right)\cos\alpha \quad ....(iii)$$

If the links PQ and PM are taken equal, QPM is an isosceles slider-crank mechanism. Again, if mechanism dimensions are so chosen that force P is perpendicular to the line of action of force F, then PQ being equal to PM, line of action of force P bisects QM. For this position $\theta = 2\alpha$.

Then taking moments of F' and P about Q,

$$F' \times (PQ)\sin 2\alpha = P(PQ)\cos\alpha$$
$$F' = (F/\cos\alpha) \quad ....(iv)$$

Hence, substituting in equation (iv),

$$\frac{F}{\cos\alpha} \times (PQ)2\sin\alpha\cos\alpha = P(PQ)\cos\alpha$$

The simplifies to, $\left(\dfrac{F}{P}\right)2\tan\alpha = 1$ or $(F/P) = \dfrac{1}{2\tan\alpha}$

Many variations of this mechanism are used in stone crushing mechanism, presses, pneumatic riveters, clutches and other applications in which one needs to develop large force out of a small force.

2.4.1 Pantograph

It is a mechanism to produce the path traced out by a point on enlarged or reduced scale. The figure (a), shows the line diagram of a pantograph in which AB = CD, BC= AD and ABCD is always a parallelogram. OQP is a straight line. P describes a path similar to that described by Q. It is used as a copying mechanism.

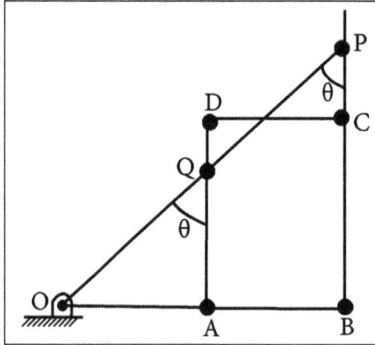

(a) Original position.   (b) Displaced position.

Proof: To prove that the path described by P is similar to that described by Q, consider Δs OAQ and OBP which are similar, because ∟BOP is common.

$\underline{|AQO}=\underline{|BPO}$ being corresponding angles as AQ∥ BP.

$$\frac{OA}{OB}=\frac{OQ}{OP}=\frac{AQ}{BP} \quad ...(i)$$

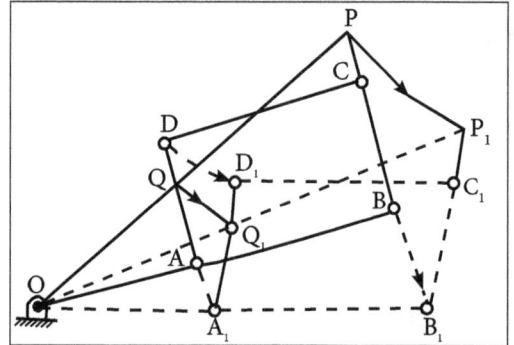

In the displaced position shown in the figure (a), as all inks are rigid,

$B_1O = BO, D_1A_1 = DA, A_1O = AO$

$P_1B_1 = PB, B_1A_1 = BA, A_1Q_1 = AQ$

Hence, $\dfrac{OA_1}{OB_1}=\dfrac{A_1Q_1}{B_1P_1}$

As $A_1B_1C_1D_1$ is a parallelogram, $A_1D_1 \parallel B_1C_1$, i.e., $A_1Q_1 \parallel B_1P_1$

OQ_1P_1 is again a straight line so that Δs OA_1Q_1 and OB_1P_1 are similar.

$$\frac{OA_1}{OB_1}=\frac{OQ_1}{OP_1}$$

From equation (i) and (ii), we get,

$$\frac{OQ}{OP}=\frac{OQ_1}{OP_1} \quad \text{becouse } OA = OA_1 \text{ and } OB = OB_1.$$

Hence QQ_1, is similar to PP_1, or they are parallel. Thus path traced by P is similar to that of Q. The pantograph is used in geometrical instruments, manufacture of irregular objects, to guide cutting tools and as indicator rig for cross head.

2.4.2 Ackerman Steering Gear Mechanism

This steering mechanism consists of a four-bar mechanism involving turning pairs only. The two opposite links AC and KL, of unequal lengths, are parallel in normal position when the vehicle moves straight.

The other two links AK and CL, which are the arms of bell-crank levers, are of equal lengths. These arms are inclined at an angle of α to the longitudinal axis when the vehicle is moving straight. Under this condition, the four-bar mechanism AKCL constitutes a trapezium with parallel and unequal sides AC and KL.

In order to steer the car to, say right, the link CL is turned clockwise so as to increase angle α. This result in c.c.w. rotation of arm AK, reducing angle α at the other end. For a given value of ratio (AK/AC) and angle α, a unique value of angle θ may be obtained for a given value φ either graphically or analytically. From the geometry of the figure.

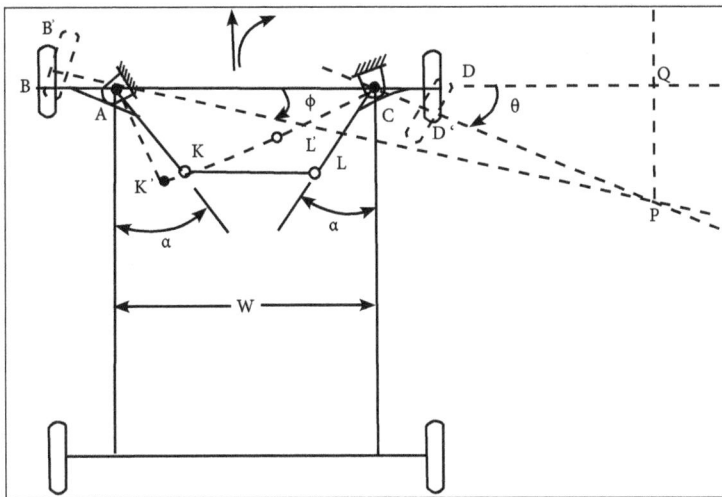

Plan view of Ackerman steering gear.

$$\cot\phi - \cot\theta = \frac{AQ - CQ}{PQ}$$

$$\left(\cot\phi - \cot\theta\right) = \left(\frac{AC}{PQ}\right) \qquad ...(i)$$

Where,

P is the point of intersection of stub-axles B'A and CD' produced. Clearly, for correct steering, the point of intersection P of front stub axle lines should lay on rear axle line produced if necessary. In other words, equation (i) should conform with,

$$\left(\cot\phi - \cot\theta\right) = \frac{W}{H}$$

For private cars with AC: AK = 8.5: 1 and α= 18°, if the distance AC is 0.4 times the wheel-base. Ackerman steering mechanism would give correct steering for θ≈24°. Thus with above proportions.

Ackerman steering gear provides correct steering in the following three positions only:

- When moving straight, i.e., when θ = 0°.

- When turning to the left at θ≈ 24°.

- When turning to the right at θ≈ 24°.

At all other values of angle θ, therefore, wheels tend to follow path along circular arcs that do not have a common centre. In all such cases, the instantaneous centre I does not lie on the axis of rear wheels but lies on a line parallel to the rear axle towards the front side.

It follows that some amount of skidding is bound to occur at steering angles other than θ = 0 and 24°. Hence, angle α and the proportions of the mechanism should be so chosen as to reduce the inherent tendency to skid to a minimum, especially for turning radii that could be followed at high speeds, where the centrifugal action also contributes to skidding.

Despite the limitation stated above, Ackerman steering gear is still the most commonly used steering gear of the two. The advantage of the Ackerman steering gear lies in the use of revolute (pin) joint rather than the sliding pairs. Resulting simplicity facilitates in up keeping. Further, as will be shown later, resulting deviation from condition of correct steering is small in the case of Ackerman steering mechanism.

The position of cross-arm (i.e., the arm of bell-crank lever) corresponding to correct angle $θ_c$ of steering, relative to the maximum angle of steering are governed by the design requirements. Smaller values of angle $θ_c$ are most frequently used while large angles occur infrequently.

It is usual therefore to select correct steering angle as equal to one-half to two-thirds of the maximum angle of steering, i.e., between 20° and 25° approximately. Angle α, needed for providing specified values angle $θ_c$ of correct steering, is established below.

Required value of A and neglecting obliquity of cress-link:

Let θ and Φ denote angles of steering corresponding to correct steering position. Now, for satisfactory operation of Ackerman steering gear, movement in a direction parallel to AC of ends K and L of cross-bar KL must always be same. Thus for displacements shown in the figure,

Projection of arc LL' on AC = Projection of arc KK' on AC

Treating arcs LL' and KK' to be equal to chords, for small angles θ and φ, we have:

$$CL\left[\sin(\alpha+\theta)\sin\alpha\right]=AK\left[\sin\alpha-\sin(\alpha-\Phi)\right]$$

As, AK = CL, simplifying further,

$$\left(\sin\alpha\cos\theta+\cos\alpha\sin\theta\right)-\sin\alpha=\sin\alpha-\left(\sin\alpha\cos\Phi-\cos\alpha\sin\Phi\right)$$

Or, $\sin\alpha\left(\cos\theta+\cos\Phi-2\right)=\cos\alpha\left(\sin\Phi-\sin\theta\right)$

$$\tan\alpha=\frac{\sin\phi-\sin\theta}{\left(\cos\theta+\cos\Phi-2\right)}\qquad...(ii)$$

Where, θ and ϕ are steering angles for the two wheels in correct steering position.

Unlike Davis' steering gear, thus, the expression for a is not quite simple. Thus to obtain desired value of α, one needs to know about the dimensions, like wheel base H and distance between pivots W of a 4-wheeler. With (W/H) ratio thus known, value of correct angle of steering ϕ for outer wheel, corresponding to each arbitrarily assumed values of angle θ for inner wheel, can then be calculated from,

$$\cot\phi-\cot\theta=\frac{W}{H}$$

With the set of values θ and ϕ for correct steering established, corresponding value of required angle α can then be calculated from equation (ii).

Problems

1. In a Davis steering gear, the distance between the pivots of front axle is 1 metre and the wheel base is 2.5 metre. Let us determine the inclination of the track α to the longitudinal axis of the car when it is moving along a straight path.

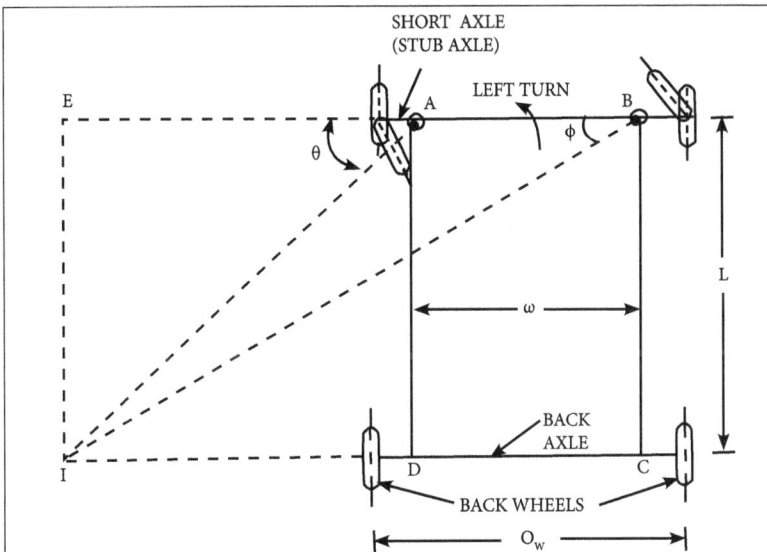

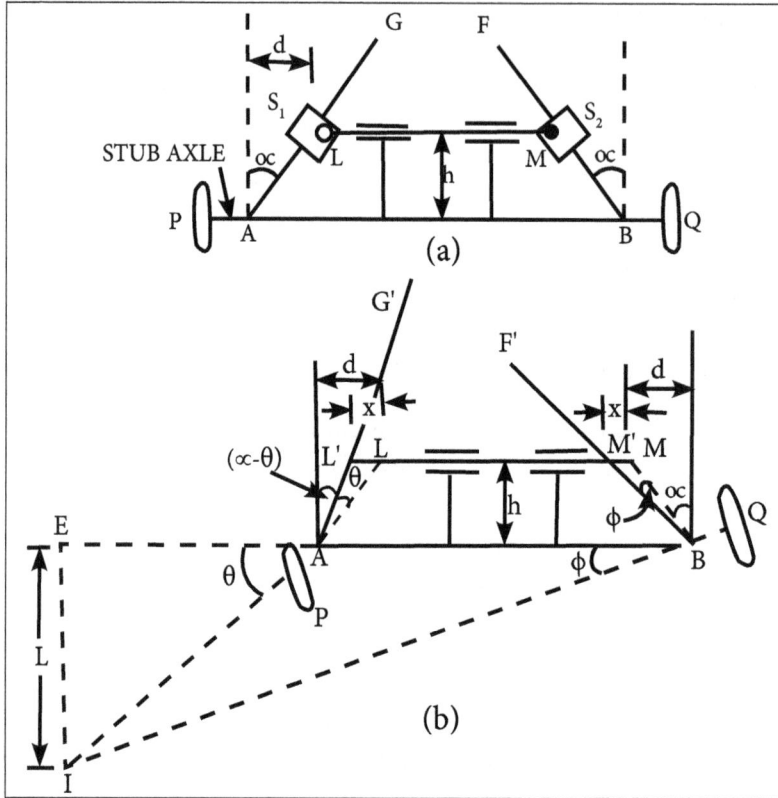

(a)

(b)

Solution:

Given:

$W = 1m = 10cm$

$H = $ wheels base $= 2.5m = 250cm$

For correct steering,

$$\cot\phi - \cot\theta = \frac{W}{H}$$

Which, leads to the condition for Davis gear as:

$$\tan\alpha = \frac{W}{2H}$$

$$= \frac{100}{(250)2} = \frac{1}{5} = 0.2$$

Velocity and Acceleration Analysis of Mechanisms (Graphical Methods)

3.1 Four Bar Mechanism

The following points are to be considered while solving problems by this method:

- Draw the configuration design to a suitable scale.
- Locate all fixed point in a mechanism as a common point in velocity diagram.
- Choose a suitable scale for the vector diagram velocity.
- The velocity vector of each rotating link is r to the link.
- Velocity of each link in mechanism has both magnitude and direction. Start from a point whose magnitude and direction is known.
- The points of the velocity diagram are indicated by small letters.

Velocity in Four bar Mechanism

The figure shows a four bar mechanism, which consists of a fixed link AD, two movable links AB and CD rotating about points A and D respectively and a connecting link BC (which is also known as coupler BC). Let the link AB is rotating at a uniform angular velocity and it is required to find the corresponding motions of the two links BC and CD.

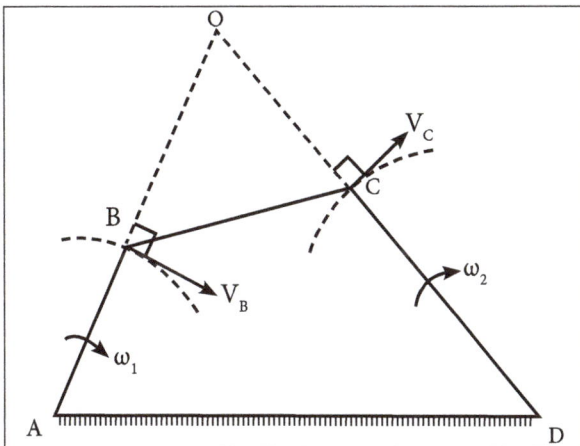

Let,

ω_{CD} = Angular velocity of link AB, rotating about A in the clock-wise direction. The value of ω_{AB} is given as ω_1.

ω_{CD} = Angular velocity of link CD, rotating about D. The value of ω_{CD} is to be calculated which is ω_2.

ω_{AB} = Angular velocity of the link ωBC.

VB =Linear velocity of point B in the direction perpendicular to AB.

= ω_{AB} x AB _____(i)

Vc =Linear velocity of point C in the direction perpendicular to CD.

= ω_{CD} x CD _____(ii)

The link AB and link CD are having motion of rotation, whereas the link BC is having motion of translation as well as rotation. The instantaneous centre of link BC for the given position is determined by drawing normal to the directions of velocity VB and VC The normal to the direction VB is the line AB whereas the normal to the direction V_c is line CD.

Hence produce line AB and CD. The intersection of these lines gives the point o (i.e., instantaneous centre) for the link BC.

The linear velocity at B is given by,

VB = ω_{BC} x BO _____(iii)

And linear velocity at C is given by,

VC = ω_{BC} x CO _____(iv)

Equating equations (i) and (iii), we get

$$\omega_{AB} \times AB = \omega_{BC} \times BO$$

$$\omega_{BC} = \frac{\omega_{AB} \times AB}{BO}$$

In equation the values ωAB are given. The value of BO is obtained from figure. Hence ω_{BC} (angular velocity of link BC) can be determined. Again equating equations (ii) and (iv), we get

$$\omega_{CD} \times CD = \omega_{BC} \times CO$$

$$\omega_{CD} = \frac{\omega_{CB} \times CO}{BO}$$

In the above equation value of CD is obtained from figure. And as well as the value of ω_{ac} and ω_{CD}.

Displacement

Displacement is a vector quantity that refers to "how far out of place an object is". It is the object's overall change in position. All particles of a body move in parallel planes and travel by same distance which is known as linear displacement and is denoted by 'x'.

A body rotating about a fixed point in such a way that all particles move in circular path and angular displacement is denoted by 'θ'.

Velocity

Rate of change of displacement is velocity. Velocity can be linear velocity or angular velocity.

Linear velocity is defined as the rate of change of linear displacement,

$$V = \frac{dx}{dt}$$

Angular velocity is defined as the rate of change of angular displacement,

$$\omega = \frac{d\theta}{dt}$$

Relation between linear velocity and angular velocity:

$$x = r\theta$$

$$\frac{dx}{dt} = r\frac{d\theta}{dt}$$

$$V = r\omega$$

$$\omega = \frac{d\theta}{dt}$$

Acceleration

The rate of change of velocity is called as acceleration.

$$f = \frac{dv}{dt} = \frac{d^2x}{dt^2}$$ Linear Acceleration (Rate of change of linear velocity).

Thirdly $\alpha = \dfrac{d\omega}{dt} = \dfrac{d^2\theta}{dt^2}$ Angular Acceleration (Rate of change of angular velocity).

Absolute Velocity

Velocity of a point with respect to a fixed point (zero velocity point).

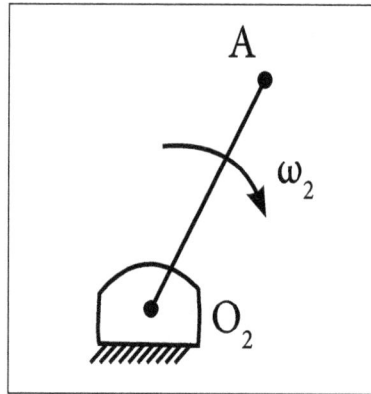

$$V_a = \omega_2 \times r$$
$$V_a = \omega_2 \times O_2A$$

Va02 - Absolute velocity.

3.1.1 Slider Crank Mechanism and Simple Mechanisms by Vector Polygons

Velocity and acceleration analysis by vector polygons: Relative velocity and accelerations of particles in a common link, relative velocity and accelerations of coincident particles on separate link, Coriolis component of acceleration.

Velocity and acceleration analysis by complex numbers. Analysis of single slider crank mechanism and four bar mechanism by loop closure equations and complex numbers.

Four bar mechanism: In a four bar chain ABCD link AD is fixed and in 15 cm long. The crank AB is 4 cm long rotates at 180 rpm (cw) while link CD rotates about D is 8 cm long BC = AD and BAD = 60°. Let us find angular velocity of link CD.

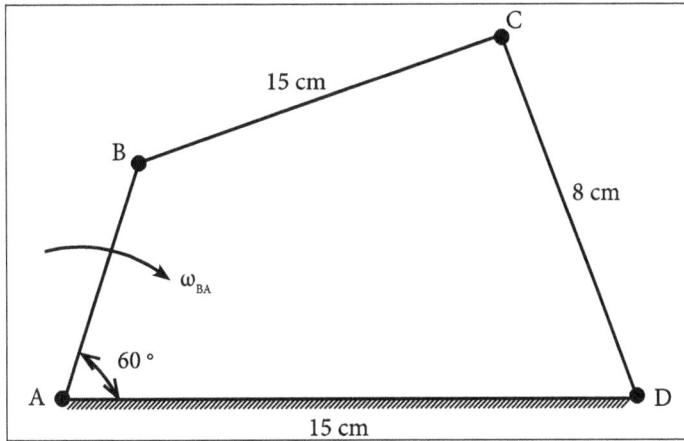

Configuration Diagram.

$$V_b = \omega r = \omega_{ba} \times AB = \frac{2\pi \times 120}{60} \times 4 = 50.24 \, \text{cm} / \text{sec}$$

Choose a suitable scale:

$$1 \, \text{cm} = 20 \, \text{m} / \text{s} = \overrightarrow{ab}$$

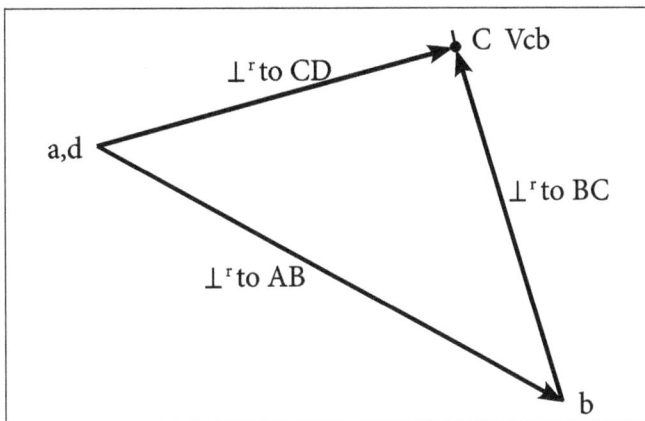

Velocity vector diagram.

$$V_{cb} = \overrightarrow{bc}$$

$$V_c = \overrightarrow{dc} = 38 \, \text{cm} / \text{sec} = V_{cd}$$

We know that, $V = \omega R$

$Vcd = \omega CD \times CD$

$$\omega_{cD} = \frac{V_{cD}}{CD} = \frac{38}{8} = 4.75 \, \text{rad} / \text{sec} \, (\text{cw})$$

Slider Crank Mechanism

In a crank and slotted lever mechanism crank rotates at 300 rpm in a counter clockwise direction. Let us find:

- Angular velocity of connecting rod.

- Velocity of slider.

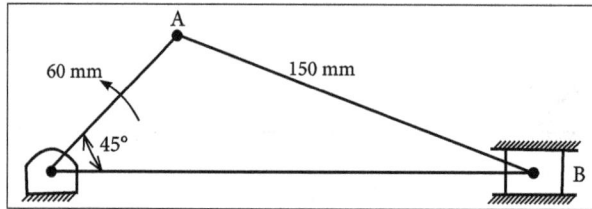

Configuration diagram.

Step 1: Determine the magnitude and velocity of point A with respect to o,

$$V_A = \omega_{O1A} \times O_2A = \frac{2\pi \times 300}{60} \times 60$$

$$= 600\pi \text{ mm / sec.}$$

Step 2: Choose a suitable scale to draw velocity vector diagram.

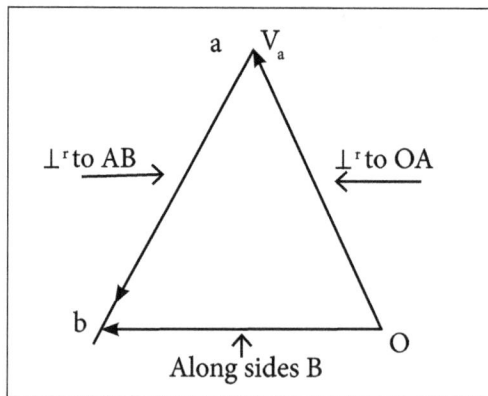

Velocity vector diagram.

$$V_{ab} = \overrightarrow{ab} = 1300 \text{ mm / sec}$$

$$\omega_{ba} = \frac{V_{ba}}{BA} = \frac{1300}{150} = 8.66 \text{ rad / sec}$$

$$V_b = \overrightarrow{ob} \text{ velocity of slider.}$$

Note: Velocity of slider is along the line of sliding.

3.1.2 Relative Velocity and Acceleration of Particles in a Common Link

Consider a rigid link PQ of length r which rotates about a fixed point P with a uniform angular velocity ω rad/s in clockwise direction as shown in the figure (a). When the link PQ takes a small turn through an angle δθ in a time δt, the point Q will travel along the arc QQ' as shown in the figure (b). The velocity of point Q relative to the fixed point P can be expressed as given by,

$$v_{qp} = \frac{\delta\theta}{\delta t} \times r \quad \text{or} \quad v_{qp} = \frac{d\theta}{dt} \times r \quad \text{when} \quad \delta t \to 0 \quad \text{or} \quad v_{qp} = \omega r$$

This shows that as the time δt approaches zero, the arc QQ' will be perpendicular to link PQ. The magnitude of velocity of point Q is ωr and its direction is perpendicular to the axis of the link in the direction of rotation. Therefore, velocity vector of link PQ can be represented as shown in the figure (c).

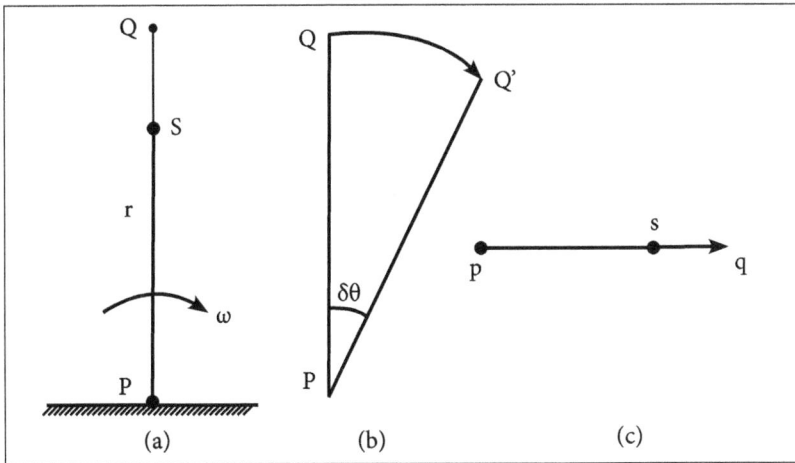

Motion of link and its velocity vector.

Consider an intermediate point S on the link PQ. The velocity of point S relative to the point P is:

$$v_{sp} = \omega \cdot SP$$

And velocity vsp can be represented by vector $\overrightarrow{ps}$ on vector diagram such that:

$$\frac{v_{sp}}{v_{qp}} = \frac{\overrightarrow{ps}}{\overrightarrow{pq}} = \frac{\omega \times SP}{\omega \times QP} = \frac{SP}{QP}$$

Hence the point s divides the vector $\overrightarrow{pq}$ in the same ratio as the point S divides the link PQ. Thus, this law of proportionality is useful in drawing the velocity polygon and finding the relative velocities of points on the link.

3.2 Relative Velocity and Accelerations of Coincident Particles on Separate Links

In geometry, two points are called coincident when they are actually the same point as each other. The same word has also been used more generally to other forms of incidence or special position between geometric objects.

When a point on one link is sliding along another rotating link, then the point is known as coincident point.

When a rotating link contains a slot along with another link constrained to slide, then a point moves with respect to another moving link. In such cases, it is not convenient to describe the absolute motion of the point. In analyzing the velocities of various machine components, of the above type, the method to be used is explained with the figure as follows:

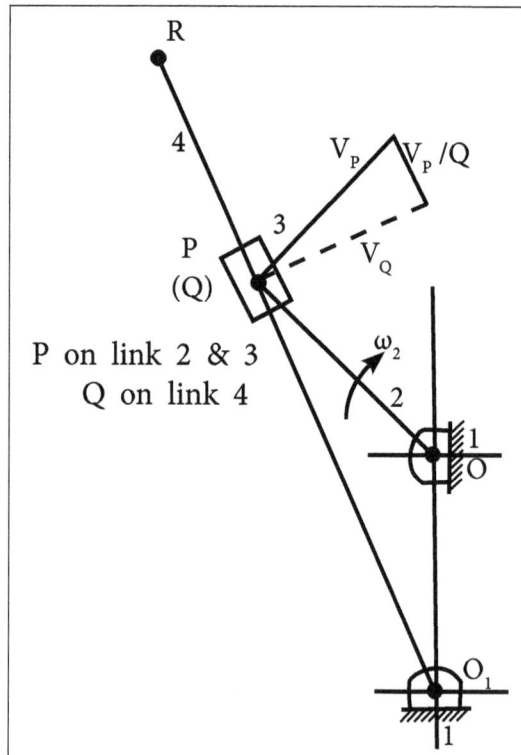

Crank and slotted lever quick return mechanism.

Figure shows a mechanism, which is a part of crank and slotted lever quick return mechanism.

The crank OP (link 2) is pivoted at O. The slider (link 3) is pin-jointed to link 2 as shown in the figure above. The slotted lever (link 4) is pivoted at O_1 and in which slider (link 3) can slide. When the link 2, rotates with uniform angular velocity ω_2 in a clockwise direction, the velocity of point P (which is at the intersection of link 2 and link 3) is V_p and perpendicular to OP as shown.

Let us consider point Q on link 4, which is coincident with point P lying on link 3. Thus points P and Q are on the separate two links, but coincident.

When the link OP (link 2) moves in a clockwise direction about O, then the slider (link 3) moves outwards in a slotted link (link 4) and hence, the slotted link (link 4) oscillates about O_1.

$\quad V_p$ = Velocity of P,

The direction of Vp is perpendicular to OP.

$\quad V_{Q/P}$ = Relative velocity Q (on link 4) with respect to P (on link 3) which is a sliding velocity because there is a sliding pair between link 3 and link 4. The direction of $V_{Q/P}$ is along slot i.e. along link 4.

$\quad V_Q$ = Velocity of Q,

The direction of V_Q is perpendicular to O_1Q. (i.e. perpendicular to link 4) The relative velocity equation for coincident points Q and P is given by,

$\quad V_Q = V_P + V_{Q/P}$ Or (oq) = (op) + (pq) ...(i)

Magnitude of V_Q and $V_{Q/P}$ (sliding velocity) can be found out from equation (i).

3.2.1 Coriolis Component of Acceleration

When a point on one link is sliding along another rotating link, such as in quick return motion mechanism, then the coriolis component of the acceleration must be calculated.

Consider a link OA and a slider Base shown in figure (a). The slider B moves along the link OA. The point C is the coincident point on the link OA.

Let,

$\quad \omega$ = Angular velocity of the link OA at time t seconds.

$\quad v$ = Velocity of the slider B along the link OA at time t seconds.

$\quad \omega.r$ = Velocity of the slider B with respect to O (perpendicular to the link OA).

At time t seconds,

$$\left(\omega + \delta_\omega\right), \left(v + \delta_v\right) \text{and} \left(\omega + \delta_\omega\right)\left(r + \delta_r\right)$$

Corresponding values at time $\left(t + \delta_t\right)$ seconds.

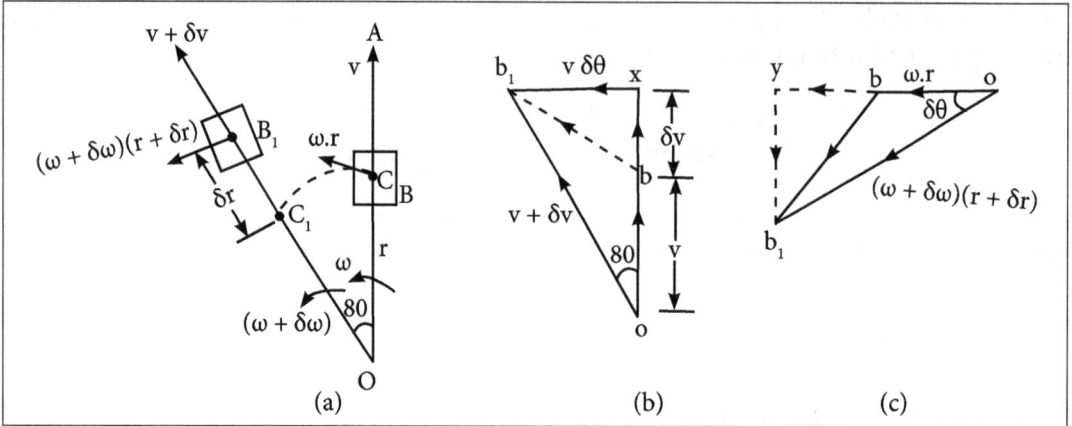

Coriolis component of acceleration.

Let us find out the acceleration of the slider B with respect to O and with respect to its coincident point C lying on the link OA.

The figure (b), shows the velocity diagram when their velocities v and $(v + \delta_v)$ are considered. In this diagram, the vector bb1 represents the change in velocity in time δ_t sec.

The vector bx represents the component of change of velocity bb1 along OA and vector xb1 represents the component of change of velocity bb1 in a direction perpendicular to OA.

Therefore,

$$bx = ox - ob = (v + \delta_v)\cos \delta_\theta - v \uparrow$$

Since, $\delta\theta$ is very small, therefore substituting $\cos\delta\theta = 1$, we have

$$bx = (v + \delta_v - v) \uparrow = \delta_v \uparrow$$

And,

$$xb_1 = (v + \delta_v)\sin \delta_\theta$$

Since, δ_θ is very small, therefore substituting $\sin \delta_\theta = \delta_\theta$, we have:

$$xb_1 = (v + \delta_v)\delta_\theta = v\delta_\theta + \delta_v.\delta_\theta$$

Neglecting $\delta_v.\delta_\theta$ being very small, therefore given by,

$$xb_1 = v \cdot \overline{\delta\theta}$$

And figure (c) shows the velocity diagram when the velocities $\omega.r$ and $(\omega + \delta_\omega)$ $(r + \delta_r)$ are considered. In this diagram, vector bb1 represents the change in velocity. Vector

yb_1 represents the component of change of velocity bb_1 along OA and vector by represents the component of change of velocity bb1in a direction perpendicular to OA. Therefore,

$$yb_1 = (\omega + \delta_\omega)(r + \delta_r)\sin \delta_\theta \downarrow$$

$$= (\omega r + \omega \delta_r + \delta_\omega r + \delta_\omega \delta_r)\sin \delta_\theta$$

Since $\delta\theta$ is very small, therefore substituting $\sin \delta\theta = \delta\theta$ in the above expression, we have:

$$yb_1 = \omega r \delta_\theta + \omega \delta_r \delta_\theta + \delta_\omega r \delta_\theta + \delta_\omega \delta_r \delta_\theta$$

$$= \omega r \delta\theta \downarrow, \text{ acting radially inwards:}$$

$$by = oy - ob = (\omega + \delta_\omega)(r + \delta_r)\cos \delta_\theta - \omega r$$

$$= (\omega r + \omega \delta_r + \delta_\omega r + \delta_\omega \delta_r) \cos \delta_\theta - \omega r$$

Since $\delta\theta$ is small, therefore substituting $\cos \delta\theta = 1$, we have:

$$by = \omega r + \omega \delta_r + \delta_\omega r + \delta_\omega \delta_r - \omega r = \omega \delta_r + \delta_\omega r + \delta_\omega \delta_r$$

Therefore, total component of change of velocity along radial direction.

$$= bx - yb_1 = (\delta_v - \omega r \delta_\theta) \uparrow$$

∴Radial component of the acceleration of the slider B with respect to O on the link OA, acting radially outwards from O to A,

$$a^r_{BO} = Lt \frac{\delta v - \omega r \cdot \delta\theta}{\delta t} = \frac{dv}{dt} - \omega \cdot r \times \frac{d\theta}{dt} = \frac{dv}{dt} - \omega^2 \cdot r \uparrow$$

Also, the total component of change of velocity along tangential direction.

$$= xb_1 + by = v \cdot \overline{\delta\theta} + (\omega \cdot \overline{\delta r} + r \cdot \delta\omega)$$

∴ Tangential component of acceleration of the slider B with respect to O on the link OA, acting perpendicular to OA and towards left.

$$a^t_{BO} = Lt \frac{v \cdot \delta\theta + (\omega \cdot \overline{\delta r} + r \cdot \delta\omega)}{\delta t} = v \frac{d\theta}{dt} + \omega \frac{dr}{dt} + r \frac{d\omega}{dt}$$

$$v \cdot \omega + \omega \cdot v + r \cdot \alpha = (2v \cdot \omega + r \cdot \alpha)$$

Now radial component of acceleration of the coincident point C with respect to O, acting in a direction from C to O.

$$a^r_{CO} = \omega^2 \cdot r \uparrow$$

And tangential component of acceleration of the coincident point C with respect to O, acting in a direction perpendicular to CO and towards left,

$$a^t_{CO} = \overline{\alpha \cdot r} \uparrow$$

Radial component of the slider B with respect to the coincident point C on the link OA, acting radially outwards,

$$a^r_{BC} = a^r_{BO} - a^r_{CO} = \left(\frac{dv}{dt} - \omega^2 \cdot r \right) - \left(-\omega^2 \cdot r \right) = \frac{dv}{dt} \uparrow$$

And tangential component of the slider B with respect to the coincident point C on the link OA acting in a direction perpendicular to OA and towards left.

$$a^t_{BC} = a^t_{BO} - a^t_{CO} = \left(2\omega \cdot v + \alpha \cdot r \right) - \alpha \cdot r = \overline{2\omega} \cdot v$$

This tangential component of acceleration of the slider B with respect to the coincident point C on the link is known as coriolis component of acceleration and is always perpendicular to the link.

∴ Coriolis component of the acceleration of B with respect of C,

$$a^c_{BC} = a^t_{BO} = 2\omega \cdot v$$

Where,

ω = Angular velocity of the link OA,

v = Velocity of slider B with respect to coincident point C.

Problems

1 A slider sliding at 100 mm/sec on a link, which is rotating at 6 rpm, is subjected to Coriolis acceleration. Let us find its magnitude.

Solution:

Given data:

V = 100 mm/s,

N = 60 rpm.

Coriolis acceleration,

$$a° = 2\omega.V$$

$$\omega = \frac{2\pi N}{60} = \frac{2\pi(60)}{60} = 6.28 \text{ rad}/s$$

$$= 2 \times 6.28 \times 100$$

$$a° = 1,256 \text{ mm}/s^2$$

2 Rocker Output - Two Positions with Angular Displacement (Function). Let us design a four bar Grash of crank-rocker speed motor input to give 45° of rocker motion with equal time forward and back, from a constant speed motor input.

Solution:

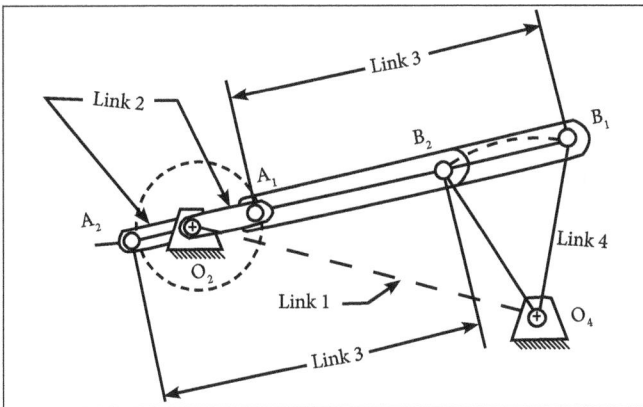

Draw the output link O_4B in both extreme positions, B_1 and B_2 in any convenient location, such that the desired angle of motion θ_4 is subtended.

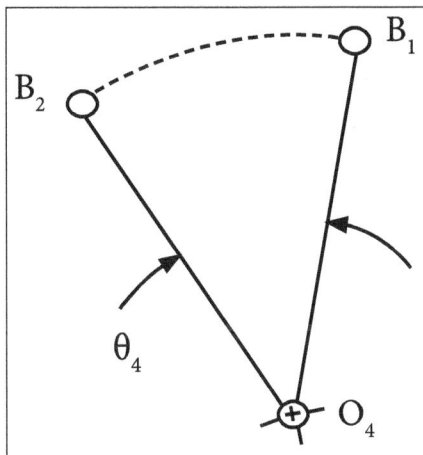

Draw the chord B_1B_2 and extended it in either direction.

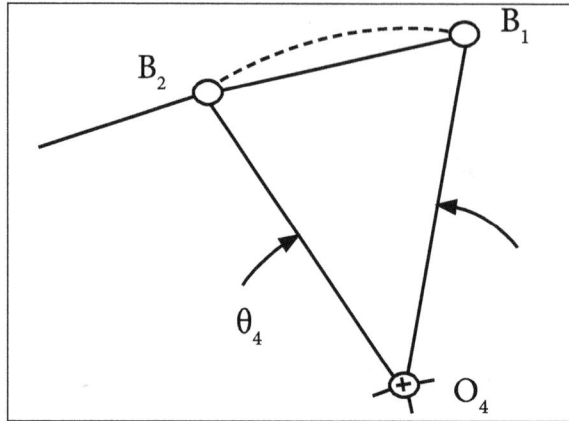

Select a convenient point O_2 on the line B_1B_2 extended.

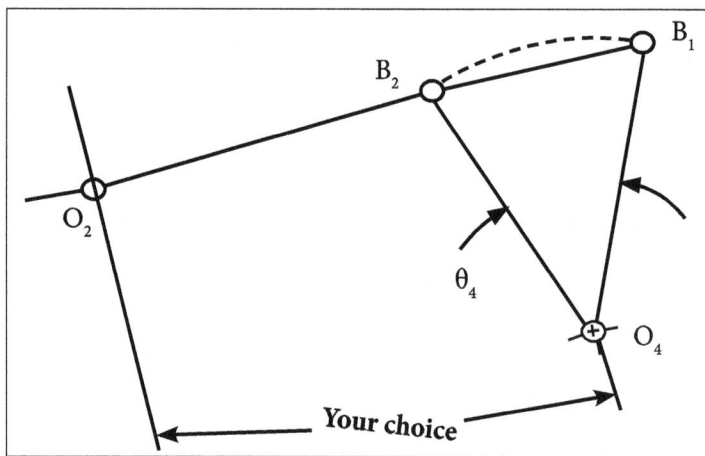

Bisect line segment B_1B_2, draw a circle of that radius about O_2.

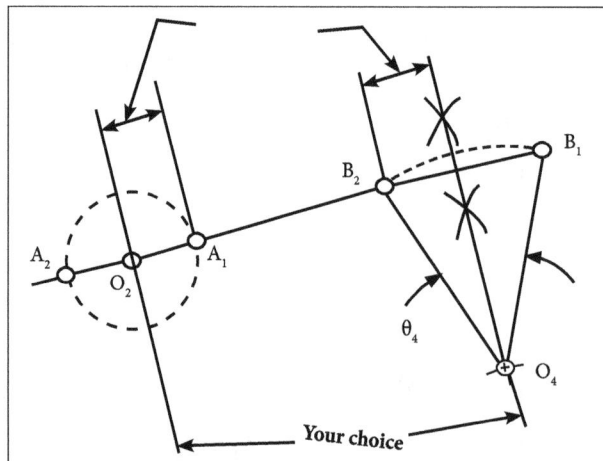

Label the two intersections of the circle and B_1B_2 extended, A_1 and A_2.

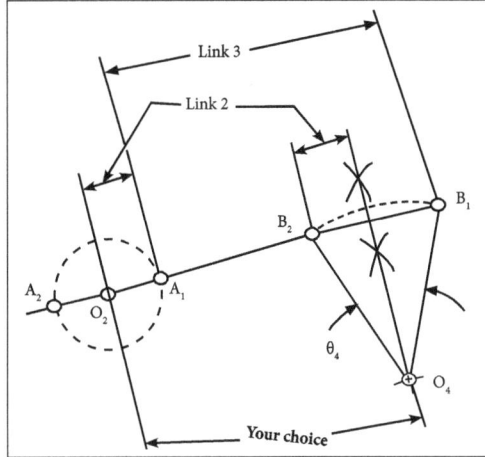

Measure the length of the coupler as A_1 to B_1 or A_2 to B_2.

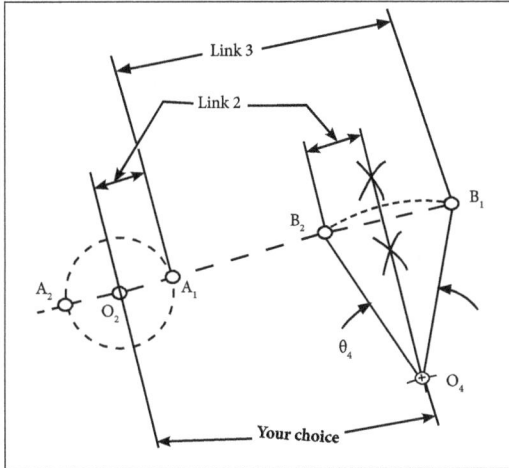

Measure ground length 1, crank length 2 and rocker length 4.

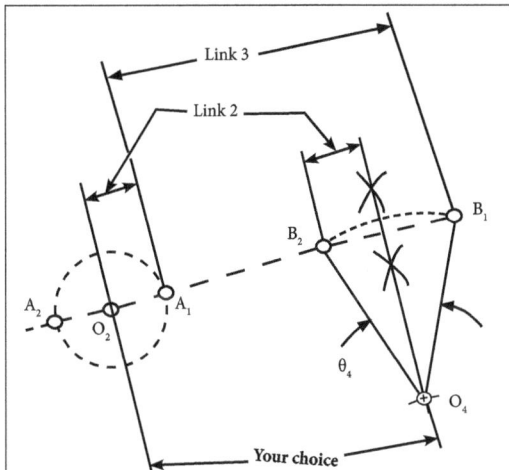

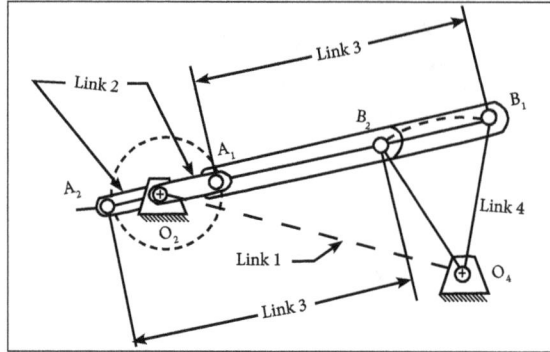

Find the Grash of condition. If non Grash of, redo steps 3 to 8 with O_2 further from O_4.

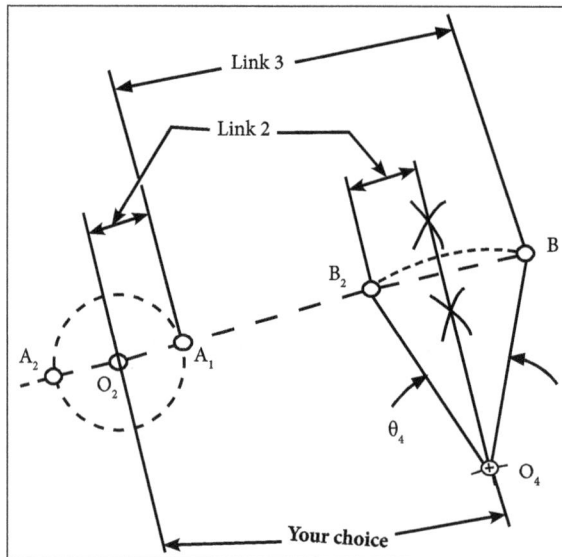

Make model of the linkage and check its function and transmission angles.

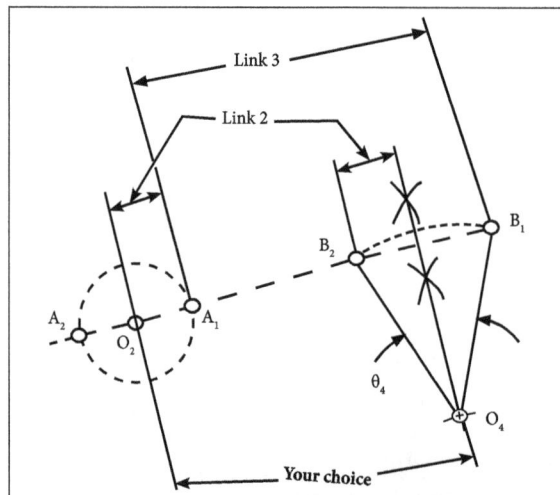

We can input the file F03-04.4br.

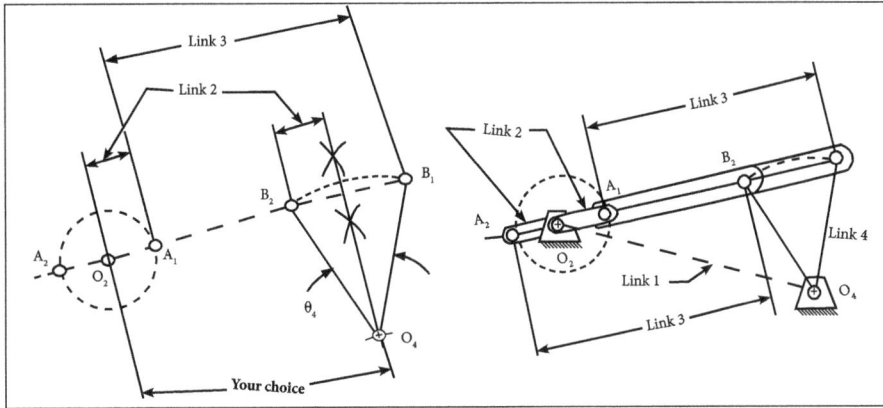

3.3   Angular Velocity and Angular Acceleration of Links

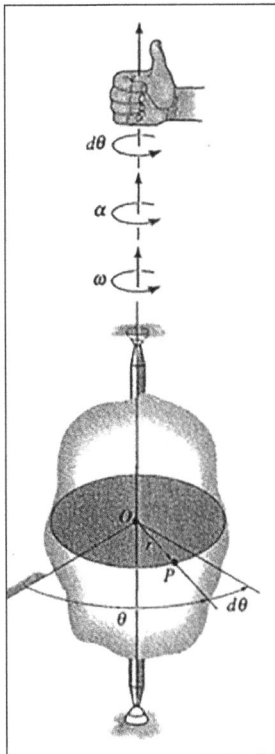

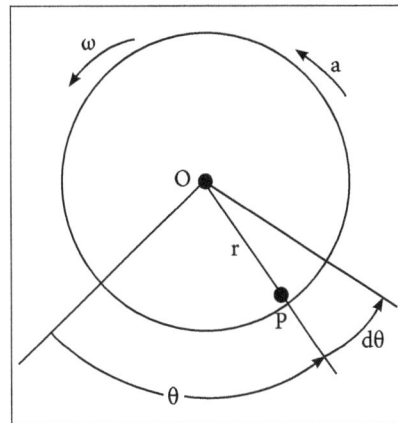

| (a) | (b) |

When a body is rotating about a fixed axis, any point P located in the body travels along a circular path. To study this motion it is first necessary to discuss the angular motion of the body about the axis.

Angular Motion

A point is without dimension and so it has no angular motion. Only lines or bodies undergo angular motion. For example, consider the body shown in figure (a) and the angular motion of a radial line r located within the shaded plane and directed from point O on the axis of rotation to point P.

Angular Position: At the instant shown, the angular position of r is defined by the angle θ, measured between a fixed reference line and r.

Angular Displacement: The change in the angular position, which can be measured as a differential $d\theta$, is called the angular displacement. This vector has a magnitude of $d\theta$, measured in degrees, radians or revolutions, where, 1 rev = 2π rad.

Since motion is about a fixed axis, the direction of $d\theta$ is always along the axis. Specifically, the direction is determined by the right hand rule. That is, the fingers of the right hand are curled with the sense of rotation, so that in this case the thumb or $d\theta$, points upward, figure(a). In two dimensions, as shown by the top view of the shaded plane, figure (b) both θ and $d\theta$ are directed counterclockwise and so the thumb points outward from the page.

Angular Velocity: The time rate of change in the angular position is called the angular velocity ω (omega). Since $d\theta$ occurs during an instant of time dt, then;

$$\omega = \frac{d\theta}{dt} \quad ...(1)$$

This vector has a magnitude which is often measured in rad/s. It is expressed here in scalar form since its direction is always along the axis of rotation, i.e., in the same direction as $d\theta$, a figure (a). When indicating the angular motion in the shaded plane, figure(b), we can refer to the sense of rotation as clockwise or counterclockwise.

Here we have arbitrarily chosen counterclockwise rotations as positive and indicated this by the curl shown in parentheses next to equation realize, however, that the directional sense of ω is actually outward from the page.

Angular Acceleration: The angular acceleration α (alpha) measures the time rate of change of the angular velocity. The magnitude of this vector may be written as,

$$\alpha = \frac{d\omega}{dt} \quad ...(2)$$

$$\alpha = \frac{d^2\theta}{dt^2} \quad ...(3)$$

The line of action of a is the same as that for ω, as shown in the figure (a) however, its sense of direction depends on whether ω is increasing or decreasing. In particular, if ω is decreasing, then α is called an angular deceleration and it therefore has a sense of direction which is opposite to ω.

By eliminating dt from equation 1 and 2, we obtain a differential relation between the angular acceleration, angular velocity and angular displacement, namely,

$$\alpha \, d\theta = \omega \, d\omega \,...(4)$$

The similarity between the differential relations for angular motion and those developed for rectilinear motion of a particle (v = ds/dt, a= dv/dt and a ds = v dv) should be apparent.

Angular Acceleration

If the angular acceleration of the body is constant, $\alpha = \alpha c$ then equation (1), (2) and (4) when integrated, yield a set of formulas which relate the body's angular velocity, angular position and time. These equations are similar to equation (4)to (6) used for rectilinear motion. The results are,

$$\omega = \omega_0 + \alpha_c t \,...(5)$$

$$\theta = \theta_0 + \omega_0 t + 1/2 \, \alpha_c t^2 ...(6)$$

$$\omega = \omega_{0^2} + 2\alpha_c (\theta - \theta_0) \,...(7)$$

Here, $\theta 0$ and $\omega 0$ are the initial values of the body's angular position and angular velocity respectively.

Problem

1. A disc having the radius R = 40 cm performs a rotation motion about its fixed center with constant angular velocity ω = 0, 5 s^{-1}. Let us determine and represent, at a given instant, the velocities and accelerations of ends of two perpendicular diameters and finally represent the distribution of the velocities on the two diameters.

Solution:

Because the disc performs a rotation motion about its center O (fixed point) the magnitudes of the velocities of points will be calculated with the relations.

$$v_A = OA.\omega = 40.0.5 = 20 \, cm/s$$
$$v_B = OB.\omega = 40.0.5 = 20 \, cm/s$$
$$v_C = OC.\omega = 40.0.5 = 20 \, cm/s$$
$$v_D = OD.\omega = 40.0.5 = 20 \, cm/s$$

As we can see all points will have velocities with the same magnitudes because they are at the same distances about the rotation center.

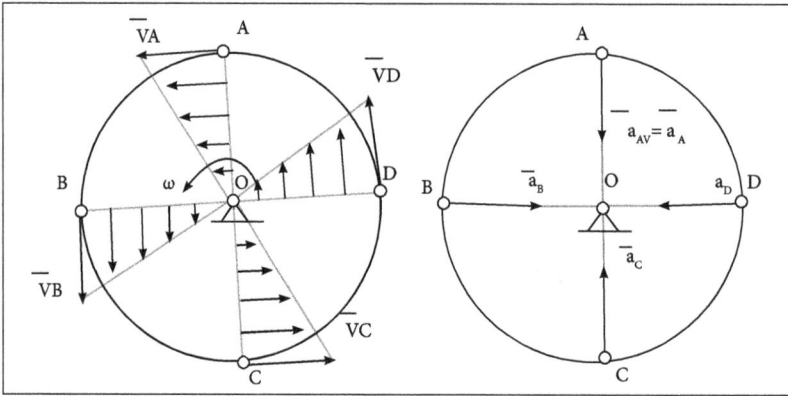

The directions of these velocities are perpendicular on the radii from point O to the given points $\overline{v_A} \perp \overline{OA}$; $\overline{v_B} \perp \overline{OB}$; $\overline{v_C} \perp \overline{OC}$; $\overline{v_D} \perp \overline{OD}$ and the senses are in the same sense of rotation, about the rotation center, as the sense of the angular velocity, namely in trigonometric sense.

Distribution of velocities on the two perpendicular diameters is linear having zero value in the rotation center O.

The acceleration of a point in the rotation motion may be determined calculating two components, one tangent and the other normal. For the point A the tangent component has the magnitude,

$$a_{Ar} = OA. \, \varepsilon = OA.\omega^2 = 0$$

Because the angular velocity is constant. The second component, normal component of the acceleration will be given by,

$$a_{Av} = OA.\omega^2 = 40.0.5^2 = 10 \text{ cm / s}^2$$

This component has the direction of the radius of the circle with the sense directed to the center O. Because all the points from the periphery of the circle are at the same distance from the center of the disc (the center of rotation) we will have the equality of the magnitudes of the accelerations of the points from the periphery,

$$a_B = a_C = a_D$$

As we have seen the accelerations may be computed directly, calculating the magnitude and directions,

$$a_A = OA \cdot \sqrt{\varepsilon^3 + \omega^4} = 10 \text{ cm / s}^2 \, ; \, tg\,\varphi = \frac{\varepsilon}{\omega^2} = 0 \, ; \rightarrow \varphi = 0.$$

3.3.1 Velocity of Rubbing

The two ends of the two links of a turning pair as shown in the figure. A pin is fixed to one of the links whereas a hole is provided in the other to fit the pin. When joined, the surface of the hole of one link will rub on the surface of the pin of the other link.

The velocity of rubbing of the two surfaces will depend upon the angular velocity of a link relative to the other.

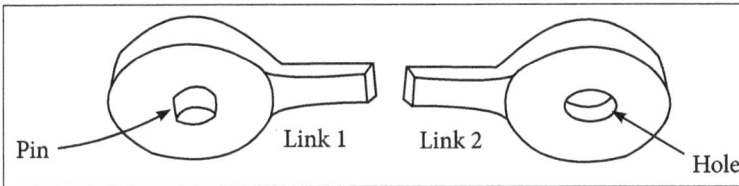

Turning pair.

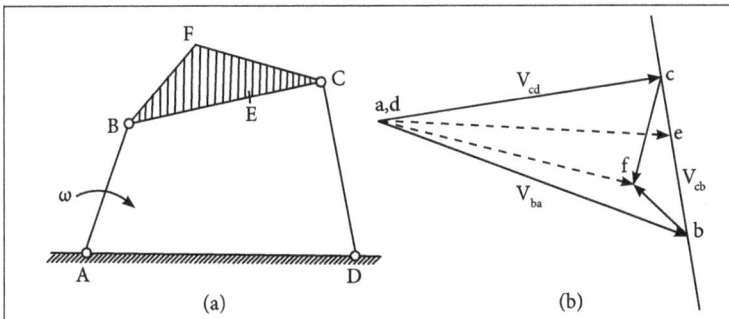

Velocity of rubbing.

Pin at A figure (a):

The pin at A joins links AD and AB. AD being fixed, the velocity of rubbing will depend upon the angular velocity of AB only.

Let r_a = Radius of the pin at A,

Then velocity of rubbing = $r_a \times \omega$.

Pin at D:

Let r_d = Radius of the pin at D,

Velocity of rubbing $= r_d . \omega_{cd}$

Pin at B:

$$\omega_{ba} = \omega_{ab} = \omega = \text{Clockwise}$$

$$\omega_{bc} = \omega_{cb} = \frac{v_{bc}}{BC} = \text{Counter} - \text{Clockwise}$$

Since the directions of the two angular velocities of links AB and BC are in the opposite directions, the angular velocity of one link relative to the other is the sum of the two velocities.

Let r_b = Radius of the pin at B,

Velocity of rubbing $= r_b \left(\omega_{ab} + \omega_{bc} \right)$.

Pin at C:

$\omega_{bc} = \omega_{cb}$ Counter-clockwise,

$\omega_{dc} = \omega_{cb}$ Clockwise,

Let r_c = Radius of the pin at C,

Velocity of rubbing $= r_c \left(\omega_{bc} + \omega_{dc} \right)$.

In this case it is found that the angular velocities of the two links joined together are in the same direction, the velocity of rubbing will be the difference of the angular velocities multiplied by the radius of the pin.

Problem

The following data refer to the dimensions of the links of a four-bar mechanism: AB = 50 mm, BC = 66 mm, CD = 56 mm and AD (fixed link) = 100 mm. At the instant when $\lfloor DAB = 60°$ the link AB has an angular velocity of 10.5 rad/s in the counter clockwise direction. Let us determine the Velocity of Point C, Velocity of Point E on the link BC while BE = 40 mm and the angular velocities of the links BC and CD and also sketch the mechanism and Indicate the data.

Solution:

Given data:

AB = 50 mm

BC = 66 mm

CD = 56 mm

AD = 100 mm

$\lfloor DAB = 60°$

ω_{BA} = 10.5 rad/sec.

$V_{BA} = \omega_{BA} \times AB = 10.5 \times 0.05 = 0.525 \ m/s$

(i) Configuration Diagram: Scale: 10 mm = 1 cm.

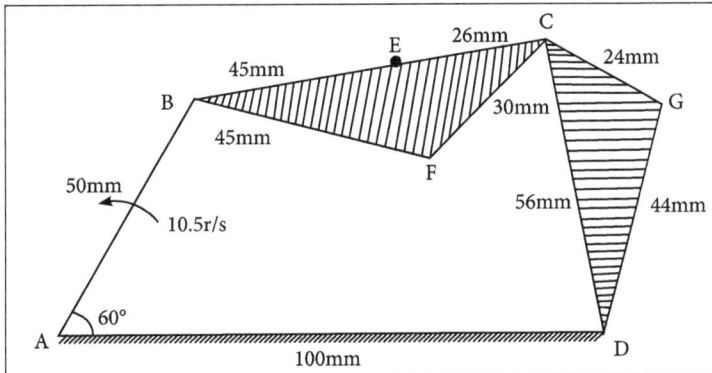

Configuration diagram.

(ii) Velocity Diagram: Scale: 0.105 m/s = 1 cm.

Directly from the velocity diagram:

v_{BA} = ab = 0.525 m/s (already known),

v_{FA} = af = 0.504 m/s,

v_{GA} = ag = 0.315 m/s.

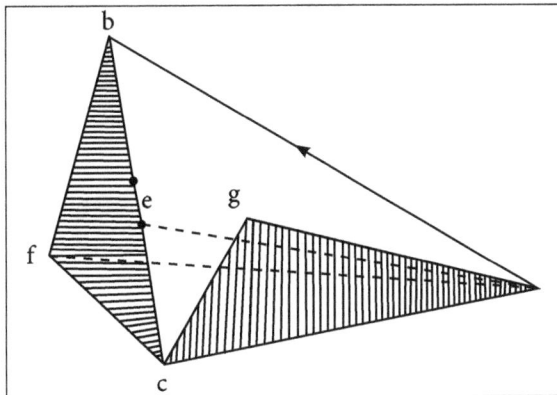

Velocity diagram.

Procedure

Configuration diagram:

- As clearly data given, we can draw ABCD.

- From point 'B' draw an arc of 45 mm radius and from C draw an arc of 30 mm radius then the arc intersecting point will be 'F'.

- From 'D', draw an arc of 44 mm radius and from C draw an arc of 24 mm radius then the arc intersecting point will be 'G'.

Velocity diagram:

- Draw ab = 0.525 m/s with anticlockwise direction at any point and perpendicular to AB.

- Since $\overline{AD}$ is fixed, the relative velocity about these points is zero, hence a and b, d will be on the same point.

- Draw a line perpendicular to $\overline{CD}$ at 'd' and draw a line perpendicular to $\overline{BS}$ at 'b' to cut each other.

- Locate the cutting point as 'c'.

- To locate the offset point F: Draw a line perpendicular to $\overline{BF}$ at 'b' and line perpendicular to $\overline{CF}$ at 'c' and locate the cutting point as 'f' and joint $\overline{af}$.

- To locate the offset point G: Draw a line perpendicular to $\overline{CV}$ at 'c' and line perpendicular to $\overline{GD}$ at 'd' and locate the cutting point as 'g'.

$$\frac{v_{be}}{v_{bc}} = \frac{BE}{BC}$$

- To locate 'E' on the velocity diagram:

$$v_{be} = v_{bc} \times \frac{BE}{BC} = 0.34 \times \frac{40}{66} = 0.206 \text{ m/s}$$

(i) Velocity of point 'C' with respect to 'a':

$$V_{ca} = 3.75 \times 0.105 = 0.393 \text{ m/s.}$$

[Θ Vca = Length 'ca'× Actual scale].

(ii) Velocity of point E with respect to B:

$$V_{EB} = 0.206 \text{ m/s.}$$

(iii) Angular velocities of link BC and CD:

$$\omega_{BC} = \frac{v_{BC}}{BC} \left[\because v_{BC} = 3.25 \times 0.105; v_{BC} = 0.341 \text{ m/s} \right]$$

$$= \frac{0.341}{0.066} = 5.17 \text{ rad/sec.}$$

$$\omega_{CD} = \frac{v_{CD}}{CD} = \frac{0.393}{0.056} \left[\because v_{CD} = 3.75 \times 0.105 = 0.393 \text{ m/s} \right]$$

$$= \frac{0.393}{0.056} = 7.01 \text{ rad/sec}$$

(iv) Velocity of offset, point F with respect to 'a':

$V_{FA} = 0.504$ m/s (Direct measurement).

(v) Velocity of offset point G with respect to 'a':

$V_{GA} = 0.315$ m/s (Direct measurement).

(vi) To calculate Rubbing Velocities:

Velocity of rubbing at pin $B = V_B = (\omega_{AB} + \omega_{BC}) \times r_B$

$= (10.5 + 5.17) \times 0.040$

$V_B = 0.6268$ m/s.

Velocity of rubbing at pin $C = V_c = (\omega_{CD} + \omega_{BC}) r_C$

$= (7.01 + 5.17) \times 0.025$

$V_{c=} 0.3045$ m/s.

Velocity of rubbing at pin $A = V_A = (\omega_{AD} + \omega_{AB}) r_A$

$= (10.5) \times 0.03$

$[\Theta \omega_{AD} = 0;\ \text{sine fixed}]$

$V_A = 0.315$ m/s.

Velocity of rubbing at pin 'D' $= (\omega_{ad} + \omega_{CD}) \times r_D$

$= (7.01) \times 0.035$

$[\Theta\ \omega_{AD} = 0]$

$V_D = 0.2454$ m/s.

Velocity Analysis by Instantaneous Center Method

4.1 Kennedy's Theorem

Statement: If three bodies have motion relative to each other, their instantaneous centers should lie in a straight line.

Arnold Kennedy Theorem of Three Centers

It states that "If three bodies have motion relative to each other, their instantaneous centers should lie in a straight line".

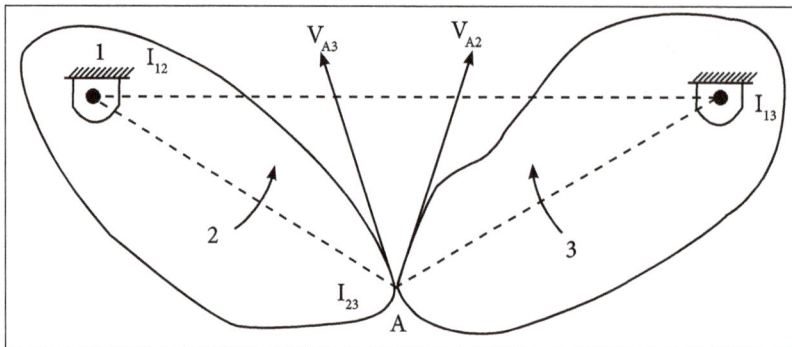

Consider a three link mechanism with link 1 being fixed link 2 rotating about I_{12} and link 3 rotating about I_{13}. Therefore I_{12} and I_{13} are the instantaneous centers for link 2 and link 3. Let us assume that instantaneous center of link 2 and 3 be at point A i.e., I_{23}. Point A is a coincident point on link 2 and link 3.

Considering A on link 2, velocity of A with respect to I_{12} will be a vector V_{A2} to link A I_{12}. Similarly for point A on link 3, velocity of A with respect to I_{13} will be r to A I_{13}. It is seen that velocity vector of V_{A2} and V_{A3} are in different directions which is impossible. Hence, the instantaneous center of the two links cannot be at the assumed position.

It can be seen that when I_{23} lies on the line joining I_{12} and I_{13} the V_{A2} and V_{A3} will be same in magnitude and direction. Hence, for the three links to be in relative motion all the three centers should lie in a same straight line.

Steps to locate instantaneous centers:

Step 1: Draw the configuration diagram.

Step 2: Identify the number of instantaneous centers by using the relation,

$$N = \frac{(n-1)n}{2}$$

Step 3: Identify the instantaneous centers by circle diagram.

Step 4: Locate all the instantaneous centers by making use of Kennedy's theorem.

4.1.1 Determination of Linear and Angular Velocity using Instantaneous Center Method Klein's Construction

Instantaneous Center

A link as a whole may be considered to be rotating about an imaginary center or about a given center at a given instant. Such a center has zero velocity; the link is at rest at this point. This is known as the instantaneous center or center of rotation.

Number of Instantaneous Centre's,

$$N = \frac{n(n-1)}{2}$$

Where, n = Number of links.

Types of Instantaneous Centers

The instantaneous centers for a mechanism are of the following three types:

- Fixed instantaneous centers.
- Permanent instantaneous centers.
- Neither fixed nor permanent instantaneous centers.

The first two types are together known as primary instantaneous centers and the third type is known as secondary instantaneous centers.

Example: Four bar mechanism, n = 4.

$$N = \frac{n(n-1)}{2} = \frac{4(4-1)}{2} = 6$$

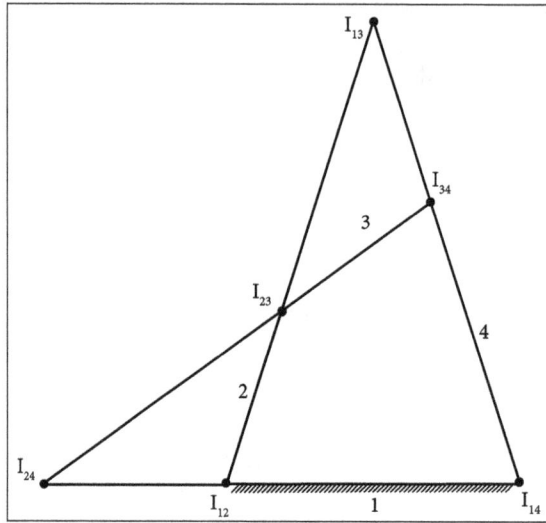

Fixed instantaneous center I_{12}, I_{14}.

Permanent instantaneous center I_{23}, I_{34}.

Neither fixed nor permanent instantaneous center I_{13}, I_{24}.

Angular Velocity using Instantaneous Center Method

Once the proper instant centers are found, they can be used to find the velocities of selected points in a rigid body. This can be done analytically. However, graphical methods are generally much faster to use. An especially useful method for finding velocities is the rotating radius method.

To develop the method, assume we have an arbitrary link moving relative to the reference system. For the sake of illustration, assume that the link is 3 and the reference link is the frame (link I). Let points P and Q be any points fixed to link 3 as shown in the figure (b).

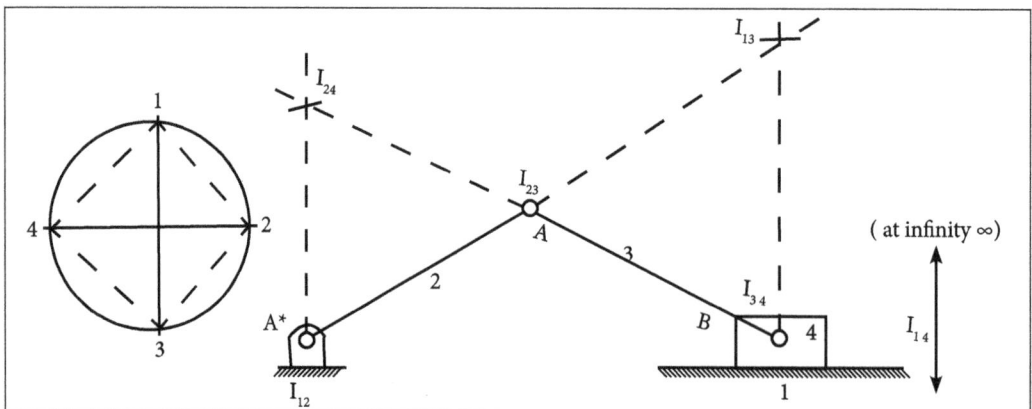

(a) Instant centers of a slider-crank linkage.

Then, we can write,

$$v_{P_3}/Q_3 = \omega_3 \times r_{P_3}/Q_3 = v_{P_3} - v_{Q_3}$$

And v_{P3}/v_{Q3} is perpendicular to the line from P to Q. If point Q_3 has zero velocity relative to link 1, then,

$$v_{P_3}/Q_3 = v_{P_3}$$

However, the only point in link 3 that has zero velocity relative to the frame is I_{13}. Therefore,

$$v_{P_3} = \omega_3 \times r_{P_3}/I_{13}$$

Because point P was any arbitrary point in link 3, this equation holds for all points in link 3. Therefore, if we know the angular velocity of the link and the instant center relative to the frame, we can compute the absolute velocity of any point in the body. Furthermore, the direction of the absolute velocity is perpendicular to the line from the point to the instant center.

For other points, only the vector r_{P3}/I_{13} will change as P changes. Considering the magnitude of the velocity,

$$\left|v_{P_3}\right| = \left|\omega_3\right|\left|r_{P_3}/I_{13}\right|$$

Because ω_3 is the same for all points in the link, the magnitude of the velocity for any other point S is given by,

$$\left|v_{S_3}\right| = \left|\omega_3\right|\left|r_{S_3}/I_{13}\right|$$

Therefore, dividing the two equations,

$$\frac{\left|v_{P_3}\right|}{\left|v_{S_3}\right|} = \frac{\left|r_{P_3}/I_{13}\right|}{\left|r_{S_3}/I_{13}\right|}$$

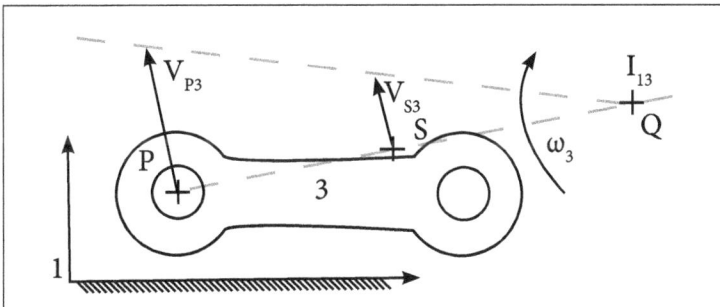

(b) Rotating radius method.

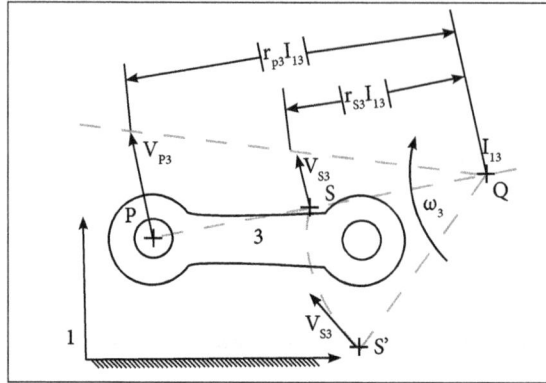

(c) Rotating radius method of obtaining the velocity of a point in a body relative to a reference frame (or another body) given the location of the instant center of the body and the velocity of some other point in the body relative to that frame.

This magnitude applies to any point that is the same distance from the instant center. The magnitude of the velocity is directly proportional to its distance from the instant center. Hence if two points in the rigid body have the same radius magnitude $\left|r_{S_3}/I_{13}\right|$, they will have the same magnitude of velocity $\left|v_{S_3}\right|$. However, the direction of their velocities will differ because the velocity is perpendicular to the line from the point to the instant center. This is illustrated by S and S' as shown in the figure (c). The actual direction of the velocity is obtained by recognizing that all points will appear to rotate about the instant center relative to the frame.

This theory is the basis for the rotating radius method. The basic procedure is to find the magnitude of the velocity of one point in the rigid body and draw that velocity vector to scale on the link. The velocity of any other point on the body then can be found by recognizing that the magnitude of the velocity is proportional to the distance from the instant center. Proportional triangles can be drawn by using the line from the original point to the instant center as a base line. Alternately, the line from the new point to the instant center can be used as a baseline.

Klein's Construction

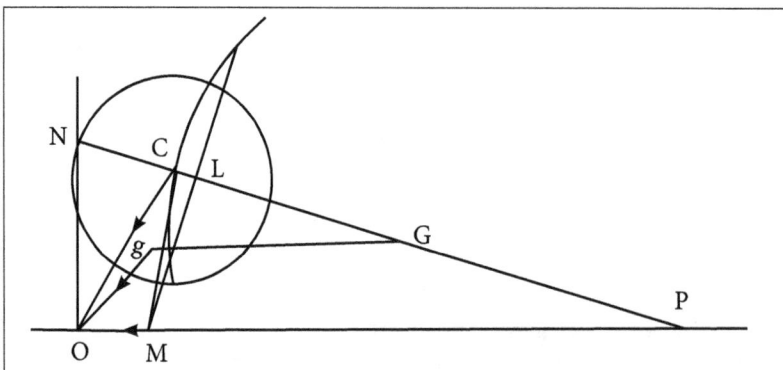

Klein's Construction for Piston Acceleration.

The above is a diagrammatic sketch of a piston, connecting rod and crank assembly where,

- PC is the connecting rod with the Crank Pin.

- OP is the line of stroke.

- OC is the crank.

- P is the gudgeon pin.

The Construction is as follows:

- Extend PC to meet the line through O perpendicular to the line of stroke. Let the point of intersection be N.

- Draw a circle with Centre C and radius CN.

- Draw a circle with CP as diameter.

- Let the common cord cut the line CP at L and the line of stroke PO at M.

- Then the quadrilateral OCLM represents, to a certain scale, the acceleration diagram for OCP. It can be shown that this scale is ω^2.

- The Centripetal acceleration of the crank pin is CO xω^2.

- The piston acceleration is MO xω^2.

- CL is the centripetal component and LM the Tangential component of the acceleration of P relative to C, so that CM is the acceleration diagram of CP.

- For any point Q on CP draw a line parallel to OP cutting CM in q. The acceleration of Q is q O xω^2 in magnitude and direction.

Problems

1. A slider crank mechanism has lengths of crank and connecting rod equal to 200 mm and 200 mm respectively locate all the instantaneous centers of the mechanism for the position of the crank when it has turned through 30_0 from IOC. Let us find velocity of slider and angular velocity of connecting rod if crank rotates at 40 rad/sec.

Solution:

Step 1: Draw configuration diagram to a suitable scale.

Step 2: Let us determine the number of links in the mechanism and find number of instantaneous.

$$N = \frac{(n-1)n}{2}$$

Centers,

n = 4 links.

$$N = \frac{4(4-1)}{2} = 6$$

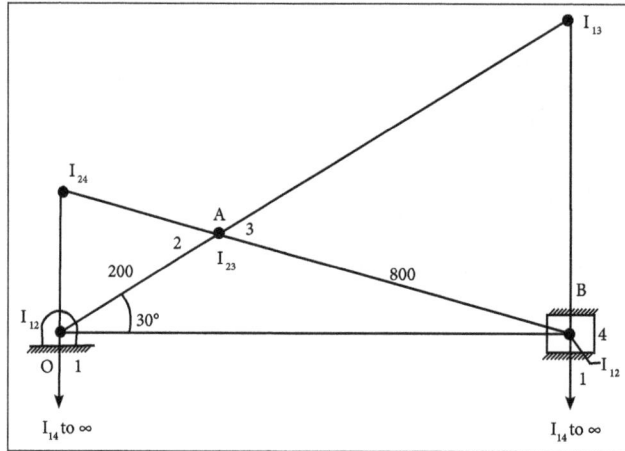

Step 3: Identify instantaneous centers.

It is a 4-bar link the resulting figure will be a square.

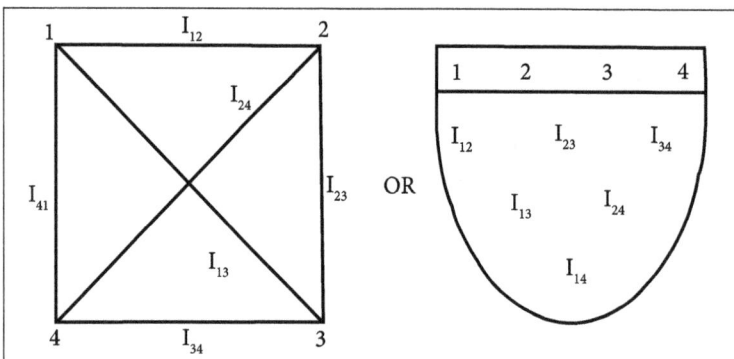

Locate fixed and permanent instantaneous centers. To locate neither fixed nor permanent instantaneous centers use Kennedy's three centers theorem.

Step 4: Velocity of different points.

$$V_a = \omega_2 AI_{12} = 40 \times 0.2 = 8 \text{ m/s}$$

Also, $V_a = \omega_3 AI_{13}$

$$\therefore \omega_3 = \frac{V_a}{AI_{13}}$$

$V_b = \omega_3 BI_{13} =$ Velocity of slider.

2. Let us locate the instantaneous centers of the slider crank mechanism shown in the figure. Let us also determine the velocity of the slider.

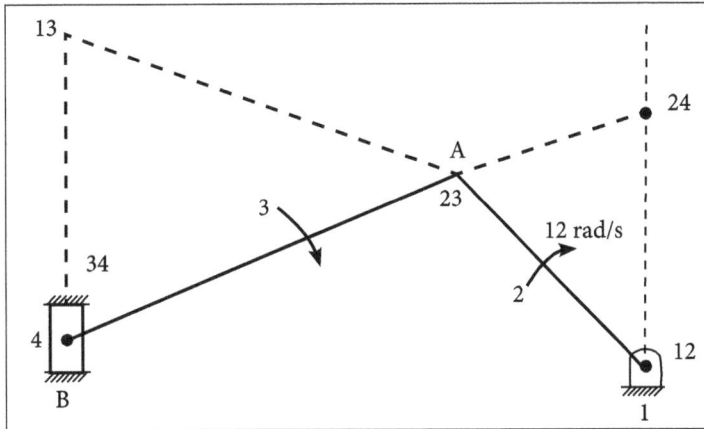

Solution:

Given:

$\omega_{oB} = 12$ rad/s,

OB = 100 mm = 0.1 m.

We know that linear velocity of the crank OB,

$V_{OA} = V_A = \omega_{oA} \times O_A$

$= 12 \times 0.10 = 1.2$ m\s.

Location of instantaneous centers,

The instantaneous centers in a slider crank mechanism are located as discussed below:

$$N = \frac{n \cdot (n-1)}{2} = \frac{4(4-1)}{2}$$

(i) Since there are four links (i.e., n = 4), therefore the number of instantaneous centers,

N = 6.

(ii) Locate the fixed and permanent instantaneous centers by inspection. These centers are I_{12}, I_{23} and I_{34} as shown in the Figure (a). Since the slides (link 4) moves on a straight surface (link 1), therefore the instantaneous center I_{14} will lie at infinite.

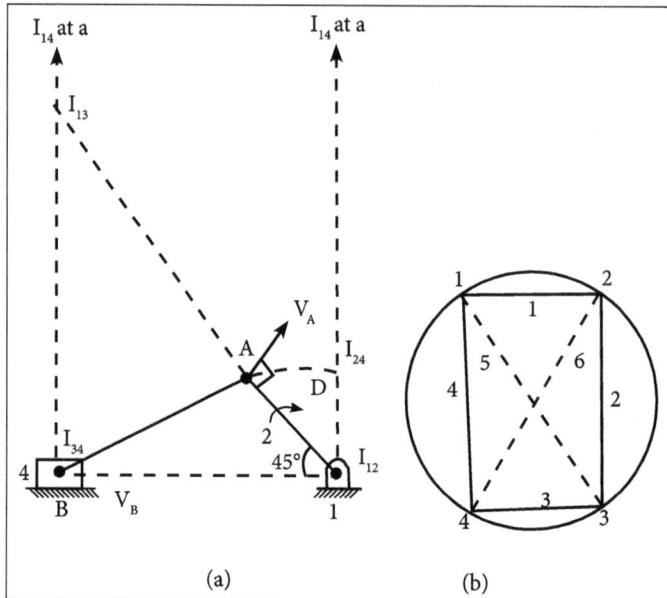

(a)　　　　(b)

(iii) Locate the other two remaining neither fixed nor permanent instantaneous centers, by

Kennedy's theorem. This is done by circle diagram as shown in the figure (b). Mark four points 1, 2, 3 and 4 (equal to the number of links in a mechanism) on the circle to indicate I_{12}, I_{23}, I_{34} and I_{14}.

(iv) Join 1 to 3 to form two triangles 123 and 341 in the circle diagram. The side 13, common to both triangles, is responsible for completing the two triangles. Therefore, the center I_{13} will lie on the intersection of $I_{12}I_{23}$ and $I_{14}I_{34}$, produced if necessary. Thus center I_{13} is located. Join 1 to 3 by a dotted line and mark number 5 on it.

(v) Join 2 to 4 by a dotted line to form two triangles 234 and 124. The side 24, common to both triangles, is responsible for completing the two triangles. Therefore the center I_{24} lies on the intersection of $I_{23}I_{34}$ and $I_{12}I_{14}$. Join 2 to 4 by a dotted line on the circle diagram and mark number 6 on it. Thus all the six instantaneous centers are located.

By measurement we find that,

$I_{13}A = 560$ mm $= 0.56$ m;

$I_{13}B = 460$ mm $= 0.46$ m;

Velocity of the slider A:

Let,

V_A = Velocity of the slider A.

We know that,

$$\frac{V_A}{I_{13}A} = \frac{V_B}{I_{13}B}$$

$$V_B = V_A \times \frac{I_{13}B}{I_{13}A}$$

$$V_B = 1.2 \times \frac{0.46}{0.56}$$

$$V_B = 0.98 \text{ m/s.}$$

4.2   Analysis of Velocity and Acceleration of Single Slider Crank Mechanism

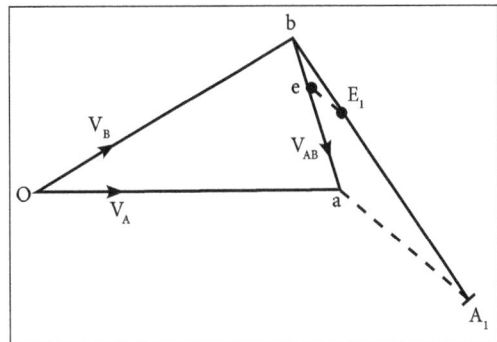

A slider crank mechanism is shown in figure (a). The slider A is attached to the connecting rod AB. Let the radius of crank OB be r and let it rotates in a clockwise direction, about the point o with uniform angular velocity ω rad/s.

Therefore, the velocity of B i.e. v_B is known in magnitude and direction. The slider reciprocates along the line of stroke AO.

The velocity of the slider A (i.e. v_A) may be determined by relative velocity method as discussed below,

1. From any point o, draw vector ob parallel to the direction of v_B (or perpendicular to OB) such that ob = v_B = ω .r, to some suitable scale, as shown in figure (b).

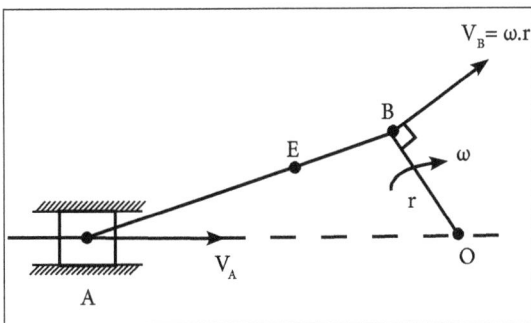

(a) Slider crank mechanism.   (b) Velocity diagram.

2. Since, AB is a rigid link; therefore the velocity of A relative to B is perpendicular to AB. Now draw vector ba perpendicular to AB to represent the velocity of A with respect to B i.e. v_{AB}.

3. From point o, draw vector oa parallel to the path of motion of the slider A (which is along AO only). The vectors ba and oa intersect at a. Now oa represents the velocity of the slider A i.e. v_A, to the scale. The angular velocity of the connecting rod AB (ω_{AB}) may be determined as follows:

$$\omega_{AB} = \frac{v_{BA}}{AB} = \frac{ab}{AB} \quad \text{(Anticlock wise about A)}$$

The direction of vector ab (or ba) determines the sense of ω_{AB} which shows that it is anticlockwise.

Note: The absolute velocity of any other point E on the connecting rod AB may also be found out by dividing vector ba such that,

be/ba = BE/BA .

This is done by drawing any line bA_1 equal in length of BA. Mark bE_1 = BE.

Join a A_1. From E_1 draw a line E_1 e parallel to aA_1.

The vector oe now represents the velocity of E and vector ae represents the velocity of E with respect to A.

Acceleration of Single Slider Crank Mechanism

A slider crank mechanism is shown in figure (a). Let the crank OB makes an angle θ with the inner dead centre (I.D.C) and rotates in a clockwise direction about the fixed point o with uniform angular velocity ω_{BO} rad/s.

Velocity of B with respect to o or velocity of B (because o is a fixed point),

$$v_{BO} = v_B = \omega_{BO} \times OB, \text{acting tengentially at B}$$

We know that centripetal or radial acceleration of B with respect to O or acceleration of B (because O is a fixed point).

$$a^r{}_{BO} = a_B = \omega^2{}_{BO} \times OB = \frac{v^2{}_{BO}}{OB}$$

Note: A point at the end of a link which moves with constant angular velocity has no tangential component of acceleration.

The acceleration diagram, as shown in figure (b), may now be drawn as discussed below:

- Draw vector o' b' parallel to BO and set off equal in magnitude of $a^r{}_{BO} = a_B$, to some suitable scale.

- From point b', draw vector b'x parallel to BA. The vector b'x represents the radial component of the acceleration of A with respect to B whose magnitude is given by,

$$a^r_{AB} = v^2_{AB} / BA$$

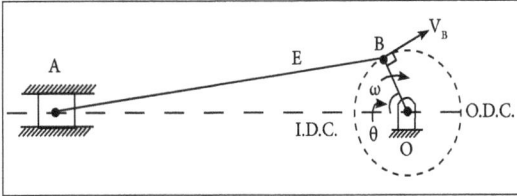

(a) Slider cranks mechanism.

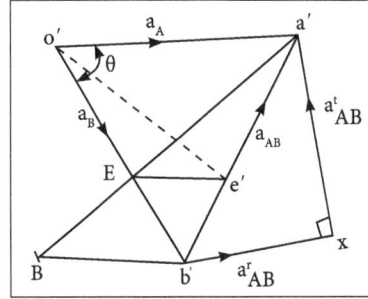

(b) Acceleration diagram.

Since the point B moves with constant angular velocity, therefore there will be no tangential component of the acceleration.

From point x, draw vector xa' perpendicular to b'x (or A B). The vector xa' represents the tangential component of the acceleration of A with respect to B i.e. a^r_{AB}.

Note: When a point moves along a straight line, it has no centripetal or radial component of the acceleration.

Since the point A reciprocates along AO, therefore the acceleration must be parallel to velocity. Therefore from o', draw o' a' parallel to AO, intersecting the vector xa' at a'. Now the acceleration of the piston or the slider A (a_A) and a^t_{AB} may be measured to the scale.

The vector b' a', which is the sum of the vectors b' x and x a', represents the total acceleration of A with respect to B i.e. a_{AB}. The vector b' a' represents the acceleration of the connecting rod AB.

The acceleration of any other point on AB such as E may be obtained by dividing the vector b' a' at e' in the same ratio as E divides AB in figure (a).

$$a'e'/a'b' = AE/AB$$

The angular acceleration of the connecting rod AB may be obtained by dividing the tangential component of the acceleration of A with respect to B (a^t_{AB}) to the length of A B. In other words, angular acceleration of A B,

$$\alpha_{AB} = a^t_{AB}/AB \text{ (Clockwise about B)}.$$

Velocity and Acceleration Analysis of Mechanisms (Analytical Methods)

5.1 Analysis of Four Bar Chain and Slider Crank Chain using Analytical Expressions

Velocity and Acceleration Analysis by Vector Polygons

Relative velocity and accelerations of particles in a common link, relative velocity and accelerations of coincident particles on separate link, Coriolis component of acceleration.

Vector Loop or Loop-Closure Equation: Four-bar linkage and slider-crank mechanism are shown using vectors as shown below,

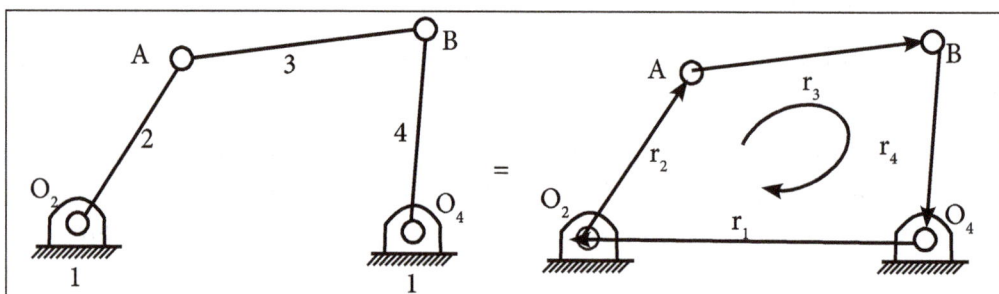

Four-Bar Linkage = Vector Loop for Four-Bar Linkage.

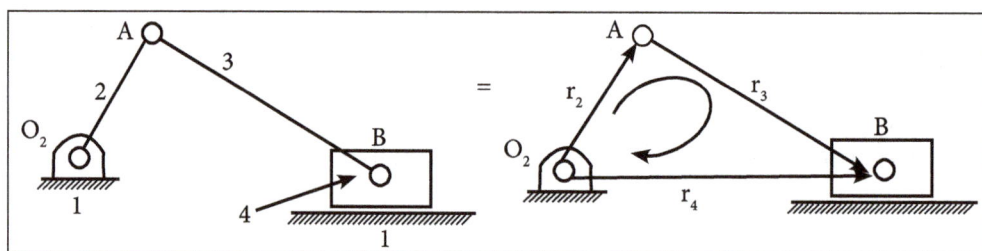

Slider-Crank Mechanism = Vector Loop for Slider-Crank Mechanism.

Vector-loop equations are:

For four-bar linkage: $r_1 + r_2 + r_3 + r_4 = 0$.

For slider-crank mechanism: $r_2 + r_3 - r_4 = 0$.

Position, velocity and acceleration analyses can be made using these vector-loop equations.

Position Analysis: The vector loop equation is written as,

$$r_1 e^{i\theta_1} + r_2 e^{i\theta_2} + r_3 e^{i\theta_3} + r_4 e^{i\theta_4} = 0 \quad ...(1)$$

Or,

$$r_1 \left(\cos\theta_1 + i\sin\theta_1 \right) + r_2 \left(\cos\theta_2 + i\sin\theta_2 \right) + r_3 \left(\cos\theta_3 + i\sin\theta_3 \right) + r_4 \left(\cos\theta_4 + i\sin\theta_4 \right) = 0$$

Collecting real and imaginary parts yields,

$$\text{Real: } r_1 \cos\theta_1 + r_2 \cos\theta_2 + r_3 \cos\theta_3 + r_4 \cos\theta_4 = 0 \quad ...(2)$$

$$\text{Image: } r_1 \sin\theta_1 + r_2 \sin\theta_2 + r_3 \sin\theta_3 + r_4 \sin\theta_4 = 0 \quad ...(3)$$

The two unknown's θ_3 and θ_4 are found using the above equations (2) and (3).

Velocity Analysis: Taking time derivative of the position vector r= re$^{I\theta}$ in general yields

$$\frac{dr}{dt} = \dot{r} = \left(\dot{r} + ir\omega \right) e^{i\theta}$$

Where, ω is known as the angular velocity of the link. Hence, taking time derivative of the vector loop equation (1) results in given below,

$$\dot{r}_1 + \left(\dot{r}_2 + ir_2\omega_2 \right) e^{i\theta_2} + \left(\dot{r}_3 + ir_3\omega_3 \right) e^{i\theta_3} + \left(\dot{r}_4 + ir_4\omega_4 \right) e^{i\theta_4} = 0 \quad ...(4)$$

Where, $\dot{r}_1 = 0$, and also $\dot{r}_2 = \dot{r}_3 = \dot{r}_4 = 0$, since r_1 is frame and the lengths of links 2, 3 and 4 are fixed. Collecting real and imaginary parts then yields,

$$\text{Real: } -r_2\omega_2 \sin\theta_2 - r_3\omega_3 \sin\theta_3 - r_4\omega_4 \sin\theta_4 = 0 \quad ...(5)$$

$$\text{Image: } r_2\omega_2 \cos\theta_2 + r_3\omega_3 \cos\theta_3 + r_4\omega_4 \cos\theta_4 = 0 \quad ...(6)$$

The problem of the velocity analysis is defined as,

Given: All r's and θ's and ω_2

Find: ω_3, ω_4

The two unknown's ω_3 and ω_4 are found using the above equations (5) and (6). Remember that a positive ω indicates counter-clock wise (ccw) rotation and a negative ω indicates clock wise (cw) rotation.

$$\frac{d\dot{r}}{dt} = \ddot{r} = \left(\ddot{r} + i2\dot{r}\omega + ir\alpha - r\omega^2 \right) e^{i\theta}$$

$$\ddot{r}_1 + \left(\ddot{r}_2 + i2\dot{r}_2\omega_2 + ir_2\alpha_2 - r_2\omega_2^2\right)e^{i\theta_2} + \left(\ddot{r}_3 + i2\dot{r}_3\omega_3 + ir_3\alpha_3 - r_3\omega_3^2\right)e^{i\theta_3}$$
$$+ \left(\ddot{r}_4 + i2\dot{r}_4\omega_4 + ir_4\alpha_4 - r_4\omega_4^2\right)e^{i\theta_4} + = 0$$

Where, $\ddot{r}_1 = 0$, and also $\dot{r}_2 = \ddot{r}_2 = \dot{r}_3 = \ddot{r}_3 = \dot{r}_4 = \ddot{r}_4 = 0$ since r_1 is frame and the lengths of links 2, 3 and 4 are fixed. Collecting real and imaginary parts then yields,

Real:

$$-r_2\alpha_2\sin\theta_2 - r_2\omega_2^2\cos\theta_2 - r_3\alpha_3\sin\theta_3 - r_3\omega_3^2\cos\theta_3 - r_4\alpha_4\sin\theta_4 - r_4\omega_4^2\cos\theta_4 = 0 \quad ...(7)$$

Image:

$$r_2\alpha_2\cos\theta_2 - r_2\omega_2^2\sin\theta_2 + r_3\alpha_3\cos\theta_3 - r_3\omega_3^2\sin\theta_3 + r_4\alpha_4\cos\theta_4 - r_4\omega_4^2\sin\theta_4 = 0 \quad ...(8)$$

The problem of the acceleration analysis is defined as,

Given: All r's, θ's and ω's and α_2

Find: α_3, α_4

The two unknown's α_3 and α_4 are found using the above equations (7) and (8). Remember that a positive α indicates ccw angular acceleration and a negative α indicates cw angular acceleration.

Slider-Crank Mechanism

Position Analysis: The vector loop equation is written as,

$$r_2e^{i\theta_2} + r_3e^{i\theta_3} - r_4e^{i\theta_4} = 0$$

Since $\theta_4 = 0$, then:

$$r_2e^{i\theta_2} + r_3e^{i\theta_3} - r_4 = 0 \quad ...(9)$$

Or,

$$r_2\left(\cos\theta_2 + i\sin\theta_2\right) + r_3\left(\cos\theta_3 + i\sin\theta_3\right) - r_4 = 0$$

Collecting real and imaginary parts yields,

$$\text{Real: } r_2\cos\theta_2 + r_3\cos\theta_3 - r_4 = 0 \quad ...(10)$$

$$\text{Image: } r_2\sin\theta_2 + r_3\sin\theta_3 = 0 \quad ...(11)$$

The problem of the position analysis is defined as,

Given: r_2, r_3 and θ_2

Find: θ_3, r_4

The two unknown's θ_3 and r_4 are found using the above equations (10) and (11).

Velocity Analysis: Taking time derivative of the vector loop equation (9) results in equation,

$$\left(\dot{r}_2 + i r_2 \omega_2\right) e^{i\theta_2} + \left(\dot{r}_3 + i r_3 \omega_3\right) e^{i\theta_3} - \dot{r}_4 = 0 \quad ...(12)$$

Where, $\dot{r}_2 = \dot{r}_3 = 0$, since the lengths of links 2 and 3 are fixed. Collecting real and imaginary parts then yields,

$$\text{Real:} \quad -r_2 \omega_2 \sin\theta_2 - r_3 \omega_3 \sin\theta_3 - \dot{r}_4 = 0 \quad ...(13)$$

$$\text{Image:} \quad r_2 \omega_2 \cos\theta_2 + r_3 \omega_3 \cos\theta_3 = 0 \quad ...(14)$$

The problem of the velocity analysis is defined as,

Given: All r's and θ's and ω_2

Find: $\omega_3, \dot{r}_4$

The velocity $\dot{r}_4$ in the equations is the velocity of the slider. The two unknowns ω_3 and $\dot{r}_4$ are found using the above equations (13) and (14). Remember that a positive $\dot{r}_4$ indicates slider moving in the direction of vector r_4.

Acceleration Analysis: Taking time derivative of the above velocity equation (12) results in equation,

$$\left(\ddot{r}_2 + i2\dot{r}_2\omega_2 + ir_2\alpha_2 - r_2\omega_2^2\right) e^{i\theta_2} + \left(\ddot{r}_3 + i2\dot{r}_3\omega_3 + ir_3\alpha_3 - r_3\omega_3^2\right) e^{i\theta_3} - \ddot{r}_4 = 0$$

Where, $\dot{r}_2 = \ddot{r}_2 = \dot{r}_3 = \ddot{r}_3 = 0$, since the lengths of links 2 and 3 are fixed. Collecting real and imaginary parts then yields.

$$\text{Real:} \quad -r_2\alpha_2\sin\theta_2 - r_2\omega_2^2\cos\theta_2 - r_3\alpha_3\sin\theta_3 - r_3\omega_3^2\cos\theta_3 - \ddot{r}_4 = 0 \quad ...(15)$$

$$\text{Image:} \quad r_2\alpha_2\cos\theta_2 - r_2\omega_2^2\sin\theta_2 + r_3\alpha_3\cos\theta_3 - r_3\omega_3^2\sin\theta_3 = 0 \quad ...(16)$$

The problem of the acceleration analysis is defined as,

Given: All r's, θ's and ω's and α_2

Find: $\alpha_3, \ddot{r}_4$

The acceleration $\ddot{r}_4$ in the equations is the acceleration of the slider. The two unknowns α_3 and $\ddot{r}_4$ are found using the above equations (15) and (16). Remember that a positive $\ddot{r}_4$ indicates slider accelerating in the direction of vector r_4.

Velocity and Acceleration Analysis by Complex Numbers

Analysis of single slider crank mechanism and four bar mechanism by loop closure equations and complex numbers. There are many ways to represent vectors. They may be defined in polar coordinates, by their magnitude and angle or in Cartesian coordinates as x and y components. These forms are of course easily convertible from one to the other using equations (i). The position vectors in the figure (a), can be represented as any of these expressions.

$$R_A = \sqrt{R_x^2 + R_Y^2} \qquad ...(i)$$

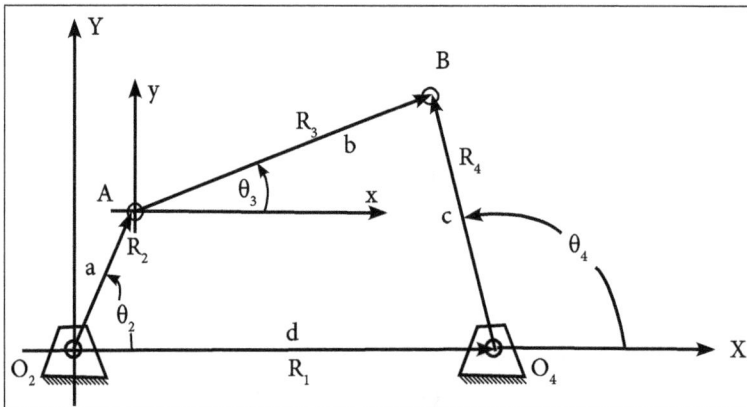

(a) Position vector for four-bar linkage.

Polar form	Cartesian form	
$R @ \angle\theta$	$r\cos\theta\,\hat{i} + r\sin\theta\,\hat{j}$	...(ii)
$re^{j\theta}$	$r\cos\theta + jr\sin\theta$	...(iii)

The equation (ii), uses unit vectors to represent the x and y vector component directions in the Cartesian form. Figure (b), shows the unit vector notation for a position vector. Equation (iii), uses complex number notation wherein the X direction component is called the real portion and the Y direction component is called the imaginary portion.

This term imaginary comes about because of the use of the notation j to represent the square root of minus one, which of course cannot be evaluated numerically. However, this imaginary number is used in a complex number as an operator, not as a value. Figure(c), shows the complex plane in which the real axis represents the X-directed component of the vector in the plane and the imaginary axis represents the Y-directed component of the same vector. So, any term in a complex number which has no j operator is an x component and a j indicates a y component.

Note in the figure (d), that each multiplication of the vector R_A by the operator j results in a counterclockwise rotation of the vector through 90 degrees. The vector $R_B = jR_A$ is directed along the positive imaginary or j axis. The vector $R_C = j^2 R_A$ is directed along the negative real axis because, $j^2 = -1$ and thus $R_C = -R_A$. In similar fashion, $R_D = j^3 R_A = -JR_A$ and this component is directed along the negative j axis.

One advantage of using this complex number notation to represent planar vectors comes from the Euler identity.

$$e^{\pm j\theta} = \cos\theta \pm j\sin\theta \qquad \text{(iv)}$$

Any two-dimensional vector can be represented by the compact polar notation on the left side of equation (iv). There is no easier function to differentiate or integrate, since it is its own derivative.

$$\frac{de^{j\theta}}{d\theta} = je^{j\theta} \qquad ...(v)$$

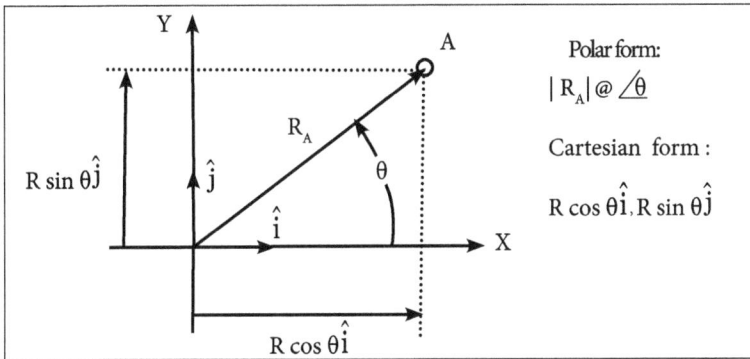

(b) Unit vector notation for the position vector.

Polar form : $R e^{j\theta}$

Cartesian form: $R\cos\theta + jR\sin\theta$

$$R = |R_A|$$

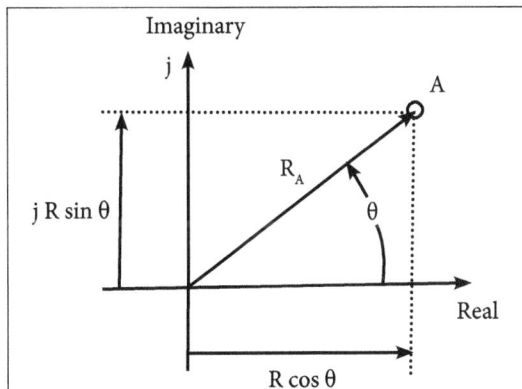

(c) Complex number representation of a position vector.

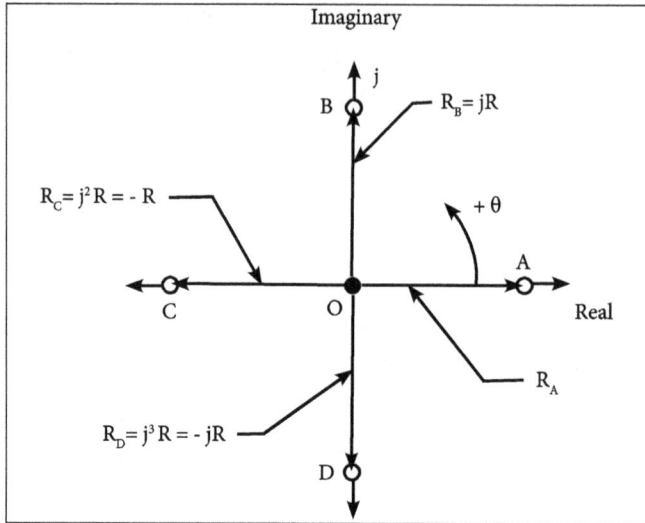

(d) Vector rotation in the complex plane.

We will use this complex number notation for vector to develop and derive the equation for position and velocity and acceleration of linkage.

Spur Gears

6.1 Gear Terminology

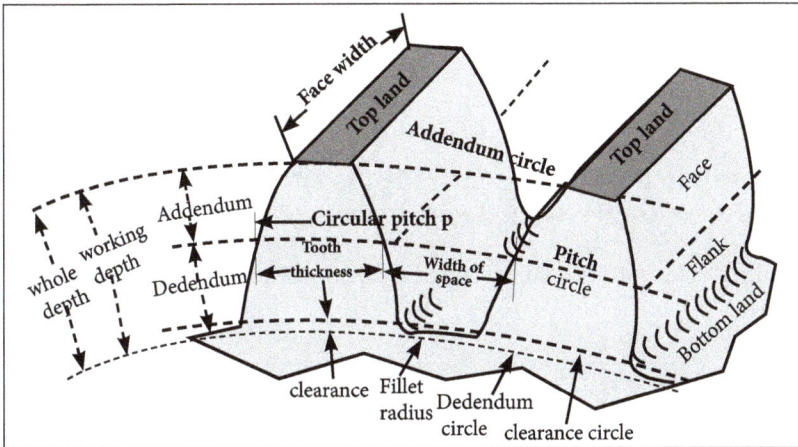

Gear tooth terminology.

- Pinion: A pinion is the smaller of two mating gears. The larger is often called the gear or the wheel.

- Pitch Circle: It is an imaginary circle which by pure rolling action would give the same motion as the actual gear.

- Pitch Circle Diameter: It is the diameter of the pitch circle. The size of the gear is usually specified by the pitch circle diameter. It is also called as pitch diameter.

- Pitch Point: It is a common point of contact between two pitch circles.

- Pitch Surface: It is the surface of the rolling discs which the meshing gears have replaced at the pitch circle.

- Pitch: Pitch of two mating gears must be same. It is defined as follows:

  ○ Circular Pitch (pc): It is the distance measured along the circumference of the pitch circle from a point on one tooth to the corresponding point on the adjacent tooth,

  Circular pitch, $P_c = \dfrac{\pi D}{T}$.

Where,

D = Diameter of pitch circle,

T = Number of teeth on the wheel.

○ Diametral Pitch (pd): It is the ratio of number of teeth to the pitch circle diameter.

Diametral pitch, $P_d = \dfrac{T}{D} = \dfrac{\pi}{P_c}$

○ Module Pitch (m): It is the ratio of the pitch circle diameter to the number of teeth,

Module, m = D/T

- Addendum Circle or Tip Circle: It is the circle drawn through the top of the teeth and is concentric with the pitch circle.

- Addendum: It is the radial distance of a tooth from the pitch circle to the top of the tooth.

- Dedendum Circle or Rout Circle: It is the circle drawn through the bottom of the circle.

- Dedendum: It is the radial distance of a tooth from the pitch circle to the bottom of the tooth.

- Clearance: It is the radial distance from the top of the tooth to the bottom of the tooth in a meshing gear. A circle passing through the top of the meshing gear is known as clearance circle.

- Total Depth: It is the radial distance between the addendum and the dedendum of a gear,

Total depth = Addendum + Dedendum.

- Working Depth: It is the radial distance from the addendum circle to the clearance circle. It is equal to the sum of the addendum of the two meshing gears.

- Tooth Thickness: It is the width of the tooth measured along the pitch circle.

- Tooth Space: It is the width of space between the two adjacent, Teeth measured along the pitch circle.

- Backlash: It is the difference between the tooth space and the tooth thickness along the pitch circle.

Backlash = Tooth space – Tooth thickness.

- Face Width: It is the width of the gear tooth measured parallel to its axis.

- Top Land: It is the surface of the top of the tooth.

- Bottom Land: The surface of the bottom of the tooth between the adjacent fillets.

- Face: Tooth surface between the pitch circle and the top land.

- Flank: Tooth surface between the pitch circle and the bottom land including fillet.

- Fillet: It is the curved portion of the tooth flank at the root circle.

- Pressure Angle or Angle of Obliquity (φ): It is the angle between the common normal to two gear teeth at the point of contact and the common tangent at the pitch point. The standard pressure angles are 14 1/2° and 20°.

- Path of Contact: It is the path traced by the point of contact of two teeth from the beginning to the end of engagement.

- Length of Path of Contact or Contact Length: It is the length of the common normal cutoff by the addendum circles of the wheel and pinion.

- Arc of Contact: It is the path traced by a point on the pitch circle from the beginning to the end of engagement of a given pair of teeth. The arc of contact consists of two parts. They are:

  ○ Arc of Approach: It is the portion of the path of contact from the beginning of the engagement to the pitch point.

  ○ Arc of Recess: It is the portion of the path of contact from the pitch point to the end of the engagement of a pair of teeth.

- Velocity Ratio: It is the ratio of speed of driving gear to the speed of the driven gear.

$$N_A/N_B = T_B/T_A$$

Where,

N_A and N_B = Speed of driver and driven,

T_a and T_B = Number of teeth on driver and driven.

- Contact Ratio: The ratio of the length of arc of contact to the circular pitch is known is contact ratio. The value gives the number of pairs of teeth in contact.

- Pressure angle: It is the angle between the common normal to two gear teeth at the point of contact and the common tangent at the pitch point. The standard pressure angles are 14 1/2° and 20°.

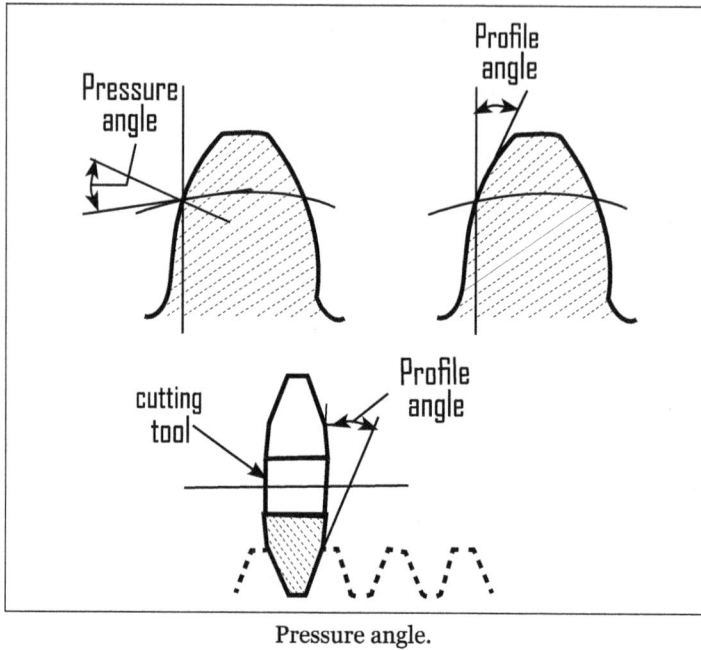

Pressure angle.

Spur Gear: Tooth Force Analysis

As shown in below figure, the normal force F will be resolved into two components, a tangential force F_t which does transmit the power and radial component F_r which does no work but tends to push the gears apart. They will hence be written as,

$$F_t = F \cos \varphi$$

$$F_t = F \sin \varphi$$

$$F_r = F_t \tan \varphi$$

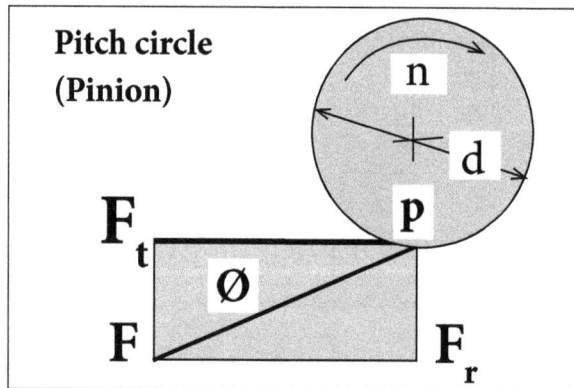

Spur Gear – Tooth Force Analysis.

The pitch line velocity V, in meters per second, is given as,

$$V = \pi dn/6000$$

$$W = F_t V / 1000$$

Where,

      d - Pitch diameter of the gear in millimeters,

      n -Rotating speed in rpm,

      W -Power in kW.

6.1.1 Law of Gearing

The common normal at the point of contact between a pair of teeth must always pass through the pitch point.

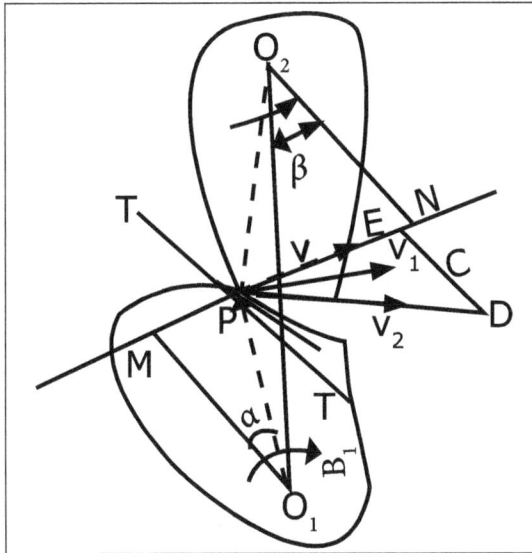

Proof

Consider the portions of the two teeth, one on the wheel 1 (pinion) and the other on the wheel 2. Let, the two teeth come in contact at point Q and the wheels rotate in the directions as shown.

TT be the common tangent and MV be the common normal to the curves at the point Q. From the centers O_1 and O_2 draw O_1M and O_2N perpendicular to MN.

Let V_1 and V_2 be the velocities on wheels 1 and 2 respectively. If the teeth remain in contact, then the components of these velocities along the common normal MN must be equal.

      $V_1 \cos \alpha = V_2 \cos \beta$

      $(\omega_1 \times O_1Q) \cos \alpha = \omega_2 \times O_2Q) \cos \beta$

$$\left(\omega_1 \times O_1 Q\right)\frac{O_1 M}{O_1 a} = \left(\omega_2 \times O_2 Q\right)\frac{O_2 N}{O_2 Q}$$

$$\frac{\omega_1}{\omega_2} = \frac{O_2 N}{O_1 M}$$

Also similar triangles $O_1 MP$ and $O_2 NP$,

$$\frac{O_2 N}{O_1 M} = \frac{O_2 P}{O_1 P}$$

$$\therefore \frac{\omega_1}{\omega_2} = \frac{O_2 N}{O_1 M} = \frac{O_2 P}{O_1 P}$$

In order to have a constant velocity ratio for all positions of the wheels, the point P must be the fixed point for the two wheels.

Velocity of Sliding

The velocity of sliding is the velocity of one tooth relative to its making tooth along the common tangent at the point of contact.

V_s = Velocity of sliding at a.

$$\frac{EC}{MQ} = \frac{V}{O_1 Q} = \omega_1$$

$$= ED - EC = \omega_1 MQ$$

$$= \omega_2 QN - \omega_1 MQ \quad \frac{EO}{QN} = \frac{V_2}{O_2 Q} = \omega_2$$

$$= \omega_2 [QP + PN] - \omega_1 [MP - QP] \quad ED = \omega_2 QN$$

$$= (\omega_1 + \omega_2) QP + \omega_2 PN - \omega_1 MP$$

Since,

$$\frac{\omega_1}{\omega_2} = \frac{O_2 P}{O_1 P} = \frac{PN}{MP} \quad [\omega_1 MP = \omega_2 PN]$$

$$V_s - (\omega_1 + \omega_2) QP$$

Problems

1. Two gear wheels mesh externally to give a velocity ratio of 3 to 1. The involute teeth

has 6 mm module and 20° pressure angle. Addendum is equal to one module. The pinion rotates at 90 rpm. Let us determine:

- Number of teeth on pinion to avoid interference and the corresponding number on the wheel;

- The length of path and arc of contact,

- Contact ratio,

- The maximum velocity of sliding.

Solution:

Given:

$G = T/t = 3$

$m = 6$ mm

$A_p = A_w = 1$ module $= 6$ mm

$\varphi = 20°$;

$N_1 = 90$ rpm

Or $\omega_1 = 2\pi \times 90160 = 9.43$ rad/s

Formula to be used:

$$t = \frac{2A_P}{\sqrt{1+G(G+2)\sin^2\phi}-1}$$

$$K_P = \sqrt{(R_A)^2 - R^2\cos^2\phi} - R\sin\phi$$

$$P_L = \sqrt{(r_A)^2 - r^2\cos^2\phi} - r\sin\phi$$

$$\text{Contact Ratio} = \frac{\text{Length of arc of contact}}{\text{Circular pitch}}$$

Number of teeth on the pinion to avoid interference with and the corresponding number of teeth on the wheel.

We know that number of teeth on the pinion to avoid interference:

$$t = \frac{2A_P}{\sqrt{1+G(G+2)\sin^2\phi}-1} = \frac{2\times 6}{\sqrt{1+3(3+2)\sin^2 20°}-1}$$

= 18.2 say 19.

And corresponding number of teeth on the wheel,

$$T = G.t = 3 \times 19 = 57$$

Length of path and arc of contact.

We know that pitch circle radius of pinion,

$$r = m \cdot \frac{t}{2} = 6 \times \frac{19}{2} = 57 \text{ mm}$$

∴ Radius of addendum circle of pinion,

$$r_A = r + \text{Addendum on pinion } (A_p)$$

$$= 57 + 6 = 63 \text{ mm and pitch circle radius of wheel,}$$

$$R = m \cdot \frac{T}{2} = 6 \times \frac{57}{2} = 171 \text{ mm}$$

∴ Radius of addendum circle of wheel,

$$R_A = R + \text{Addendum on wheel (AW)}$$

$$= 171 + 6 = 177 \text{ mm.}$$

We know that the path of approach (i.e., path of contact when engagement occurs),

$$K_P = \sqrt{(R_A)^2 - R^2 \cos^2 \phi} - R \sin \phi$$

$$= \sqrt{(177)^2 - (171)^2 \cos^2 20°} - 171 \sin 20°$$

$$= 74.2 - 58.5 = 15.7 \text{ mm}$$

And the path of recess (i.e, path of contact when, disengagement occurs),

$$P_L = \sqrt{(r_A)^2 - r^2 \cos^2 \phi} - r \sin \phi$$

$$= \sqrt{(63)^2 - (57)^2 \cos^2 20°} - 57 \sin 20°$$

$$= 33.17 - 19.5 = 13.67 \text{ mm.}$$

∴ Length of path of contact,

$$K_L = K_p + P_L = 15.7 + 13.67 = 29.37 \text{ mm}$$

We know that length of arc of contact,

$$= \frac{\text{Length of path of contact}}{\cos \phi} = \frac{29.37}{\cos 20^\circ} = 31.25 \text{ mm}$$

Contact Ratio:

$$\text{Contact Ratio} = \frac{\text{Length of arc of contact}}{\text{Circular pitch}}$$

Circular pitch = $\pi \times m$,

PC = $\pi \times 6 = 18.852$ mm,

$$\text{Contact Ratio} = \frac{31.25}{18.852} = 1.66 \text{ say } 2$$

Maximum Velocity of Sliding:

Let, ω_2 = Angular speed of wheel in rad/s

We know that,

$$\frac{\omega_1}{\omega_2} = \frac{T}{t}$$

$$\omega_2 = \omega_1 = \frac{t}{T} = 9.43 \times \frac{19}{57}$$

$$\omega_2 \frac{3}{4} \text{ rad/s}$$

∴ Maximum velocity of sliding,

$$V_s = (\omega_1 + \omega_2) KP,$$

$$= (9.43 + 3.14) 15.7.$$

$$V_s = 197.35 \text{ mm/s}.$$

2. The pressure angle of two gears is 20° and has a module of 10 mm. The number of teeth on pinion is 24 and gear on is 49. The addendum of piston and gear is same and equal to one module. Let us determine (i) the number of pairs of teeth in contact (ii) the angle of action of pinion and gear and the ratio of sliding to rolling velocity at the beginning of contact.

Solution:

Given Data:

$\emptyset = 20°,$

$m = 10 \text{ mm},$

$T_p = 24,$

$T_g = 49.$

Addendum on pinion and gear wheel = 1 module = 10 mm.

No. of pairs of teeth in contact.

$$r = \frac{m\,T_P}{2} = \frac{10 \times 24}{2} = 120 \text{ mm}$$

$$R = \frac{m\,T_G}{2} = \frac{10 \times 49}{2} = 245 \text{ mm}$$

Addendum radius of Pinion, $r_A = r + \text{addendum}$

$= 120 + 10 = 130 \text{ mm}.$

Addendum radius of gear wheel, $R_A = R + \text{Addendum}$

$= 245 + 10 = 255 \text{ mm}.$

Length of path of approved,

$$K_P = \sqrt{R_A^3 - R^2 \cos^2 \phi - R \sin \phi}$$

$$= \sqrt{(255)^2 - (255)^2 \cos^2 20 - (245 \sin 20)}$$

$$= \sqrt{65025 - 57418.5 - 83.79}$$

$$= 86.7335 \text{ mm}.$$

Length of path of reasons,

$$P_L = \sqrt{r_A^2 - r^2 \cos^2 \phi - r \, \text{sn} \, \phi}$$

$$= \sqrt{130^2 - 120^2 \cos 20 - 120 \, \text{sn} \, 20}$$

$$= \sqrt{16900 - 14400 \cos 20 - 120 \, \text{sn} \, 20}$$

$$= 57.68 \text{ mm}$$

Length of path of contact, $K_L = K_p + P_L$

$$= 66.43 + 57.68$$

$$= 144.41 \text{ mm.}$$

Length of arc of contact,

$$= \frac{\text{Length of Path of contact}}{\cos \phi}$$

$$= \frac{144.41}{\cos 20^\circ} = 153.68$$

Angle of action of pinion and gear and the ratio of sliding to rolling velocity at the beginning of contact.

Gear ratio:

$$\frac{T_G}{T_P} = \frac{49}{24} = 2.04 = 2$$

We know that,

$$\frac{\omega_P}{\omega_G} = \frac{T_G}{T_P}$$

$$\omega_G = \omega_P \times \frac{T_P}{T_G}$$

$$\omega_P \times \frac{24}{49} = 0.4897 \; \omega p$$

Rolling Velocity, $V_r = \omega_p$

$$= \omega_G \cdot R$$

$$= \omega_p \times 120$$

$$= 120 \omega_p \text{ mm/s}$$

6.2 Characteristics of Involute Action

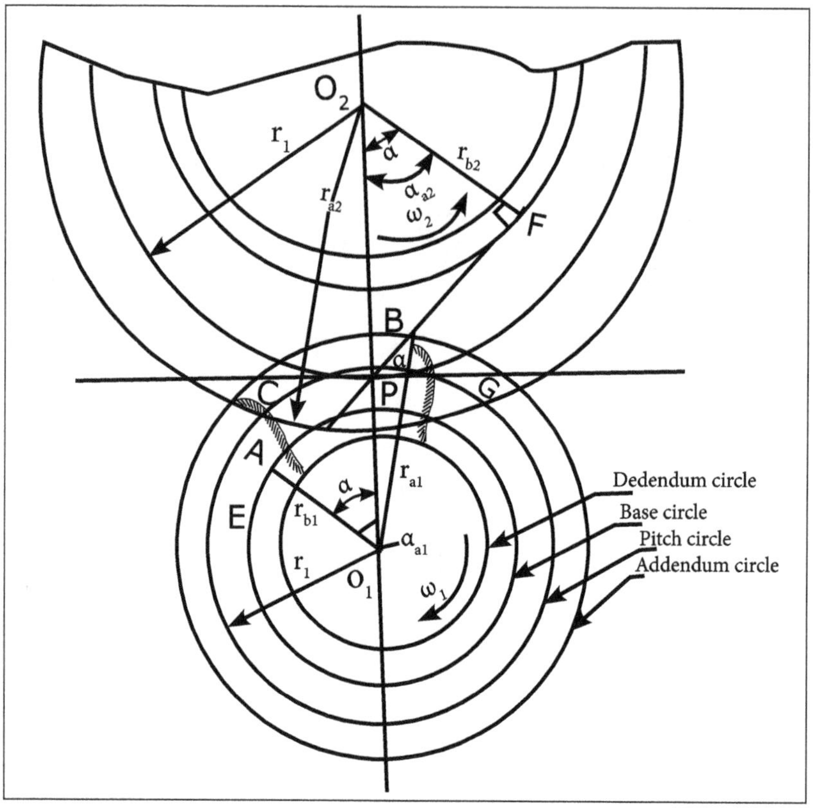

Angles of action.

The characteristics of the involute action are:

- Arc of contact.
- Length of path of contact.
- The contact ratio.

As shown in the figure below, the contact of two gear teeth begins at A and ends at B.

Addendum radius of pinion, $r_{a1} = r_1 + h_{a1}$.

Base circle radius of pinion, $r_{b1} = r_1 \cos\alpha$.

Addendum radius of gear, $r_{b2} = r_{a2} = r_2 + h_{a2}$.

Base circle radius of gear $= r_{b2}, \cos\alpha$.

Where,

r_1 = Pitch circle radius of pinion.

r_2 = Pitch circle radius of gear.

h_{a1} = Addendum of pinion.

h_{a2} = Addendum of gear.

r_{b1} = Base circle radius of pinion.

r_{b2} = Base circle radius of gear.

Length of the path recess,

$$L_r = PB = EB - EP$$

$$= \left(r^2_{a1} - r^2_{b1}\right)^{0.5} - O_1 P \sin \alpha$$

$$= \left(r^2_{a1} - r^2_{b1}\right)^{0.5} - r_1 \sin \alpha$$

Length of path of approach,

$$L_a = AP = AF - EF$$

$$= \left(r^2_{a2} - r^2_{b2}\right)^{0.5} - O_2 P \sin \alpha$$

$$= \left(r^2_{a2} - r^2_{b2}\right)^{0.5} - r_2 \sin \alpha$$

Length of path contract,

$$AB = L_p = L_r + L_a$$

$$= \left(r^2_{a1} - r^2_{b1}\right)^{0.5} + \left(r^2_{a2} - r^2_{b2}\right)^{0.5} - \left(r_1 + r_2\right) \sin \alpha$$

Length of arc contract,

$$L_c = \text{arc } CG$$

$$= \frac{AB}{\cos \alpha}$$

Maximum length of path of recess = $r_2 \sin \alpha$.

Maximum length of path of approach = $r_1 \sin \alpha$.

The contact ratio is defined as the average number of pairs of teeth, which are in contact. This can be found by nothing how many times the base pitch fits into the length of the path of contact. The contact ratio (CR) can be expressed as,

CR = length of path of contact/base pitch

$$= \frac{L_p}{P_b}$$

Where, $P_b = P\cos a = \pi m \cos a$

For a rack and a pinion,

$$L_p = \left(r^2_{a1} - r^2_{b1}\right)^{0.5} - r_1 \sin\alpha + \frac{a}{\sin\alpha}$$

Where, a = Addendum.

6.2.1 Path of Contact

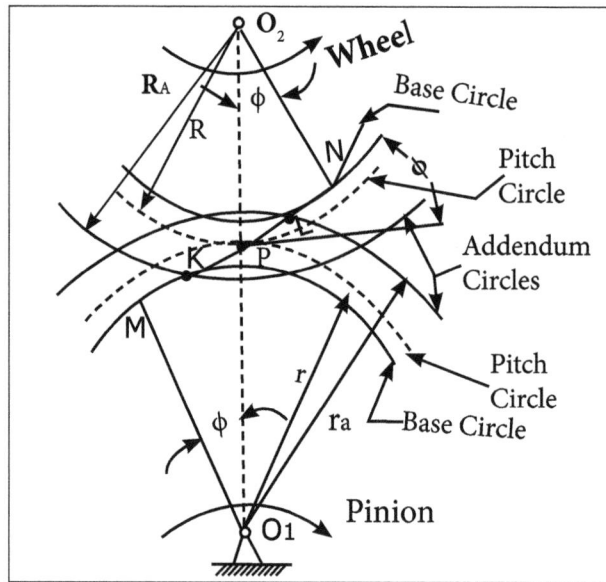

Path of Contact.

Consider a pinion driving a wheel as shown in figure. When the pinion rotates in clockwise, the contact between a pair of involute teeth begins at K (on the near the base circle of pinion or the outer end of the tooth face on the wheel) and ends at L (outer end of the tooth face on the pinion or on the flank near the base circle of wheel).

MN is the common normal at the point of contacts and the common tangent to the base circles. The point K is the intersection of the addendum circle of wheel and the common tangent. The point L is the intersection of the addendum circle of pinion and common tangent.

The length of path of contact is the length of common normal cut-off by the addendum circles of the wheel and the pinion. Thus the length of part of contact is KL which is the sum of the parts of path of contacts KP and PL. Contact length KP is called as path of approach and contact length PL is called as path of recess.

$r_a = O_1L$ = Radius of addendum circle of pinion.

$R_A = O_2K$ = Radius of addendum circle of wheel.

$r = O_1P$ = Radius of pitch circle of pinion.

$R = O_2P$ = Radius of pitch circle of wheel.

Radius of the base circle of pinion = $O_1M = O_1P \cos \varphi = r \cos \varphi$.

And,

Radius of the base circle of wheel = $o_2N = O_2P \cos \varphi = R \cos \varphi$.

From right angle triangle O_2KN,

$$KN = \sqrt{(O_2K)^2 - (O_2N)^2}$$

$$= \sqrt{(R_A)^2 - R^2 \cos^2 \phi}$$

$$PN = O_2P \sin \varphi = R \sin \varphi$$

Path of approach: KP

$$KP = KN - PN$$

$$= \sqrt{(R_A)^2 - R^2 \cos^2 \phi} - R \sin \phi$$

Similarly from right angle triangle O_1ML,

$$ML = \sqrt{(O_1L)^2 - (O_1M)^2}$$

$$= \sqrt{(r_a)^2 - r^2 \cos^2 \phi}$$

$$MP = O_1P \sin \varphi = r \sin \varphi$$

Path of recess: PL

$$PL = ML - MP$$

Length of path of contact = $= \sqrt{(r_a)^2 - r^2 \cos^2 \phi} - r \sin \phi$

$$KL = KP + PL$$

$$= \sqrt{\left(R_A\right)^2 - R^2 \cos^2 \phi} + \sqrt{\left(r_a\right)^2 - r^2 \cos^2 \phi} - \left(R + r\right) \sin \phi$$

6.2.2 Arc of Contact

Pinion with center O_1, in mesh with wheel with center O_2, MN is the common tangent to the base circles and KL is the path of contact between two mating gears.

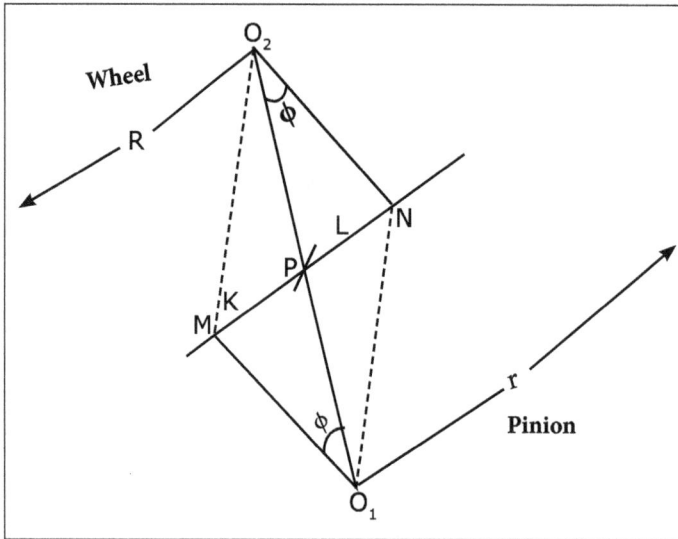

Maximum length of arc of contact.

If the radius of the addendum circle of pinion is increased to O_1N, the point of contact L will move from L to N when this radius is further increased, the point of contact L will be on the inside of base circle of wheel.

The tip of tooth on the pinion will then undercut the tooth on the wheel at the root and remove part of the in-volute profile of tooth on the wheel. This effect is known as Interference.

When interference is just avoided, the maximum length of path of contact is MN.

Maximum Length of path of approach,

$$MP = r \sin \varphi$$

Maximum Length of path of recess,

$$PN = R \sin \varphi$$

Maximum Length of path of,

$$MN = MP + PN$$

$$= (r + R) \sin \varphi$$

Maximum Length of arc of contact,

$$= \frac{\text{Maximum Length of Path of Contact}}{\cos \phi}$$

$$= \frac{(r + R) \sin \phi}{\cos \phi}$$

$$= (r + R) \tan \varphi$$

Minimum number of teeth required for pinion:

T_p = Number of teeth on the pinion,

T_G = Number of teeth on the gear wheel,

M = Module of the teeth,

γ = Pitch circle radius of the pinion = $MT_p/2$,

R = Pitch circle radius of the gear wheel = $MT_G/2$,

$$G = \text{Gear ratio} = \frac{T_G}{T_P} = \frac{1}{r},$$

φ = Pressure angles.

From triangle O_1NP,

$$(O_1N)^2 = (O_1P)^2 + (PN)^2 - 2 \times O_1P \times PN \times \cos O_1PN$$

From the above figure,

$$PN = O_2 P \sin \varphi = R \sin \varphi$$

$$(O_1N)^2 = \gamma^2 + R_2 \sin \varphi - 2\gamma \cdot R \sin \varphi \cos(90° + \varphi)$$

$$= \gamma^2 + R^2 \sin^2\varphi + 2y \cdot R \sin2 \varphi$$

$$= \gamma^2 \left[1 + \frac{R^2 \sin^2 \phi}{\gamma^2} + \frac{2R^2 \sin^2 \phi}{\gamma} \right] = \gamma^2 \left[1 + \frac{R}{\gamma} \left(\frac{R}{\gamma} + 2 \right) \sin^2 \phi \right]$$

Therefore, limiting radius of the pinion addendum circle,

$$O_1M = \gamma \sqrt{1 + \frac{R}{\gamma} (R_\gamma + 2) \sin^2 \phi}$$

$$= \frac{MT_P}{2} \sqrt{1 + \frac{T_G}{T_P}\left(\frac{T_G}{T_P} + 2\right) \sin^2 \phi}$$

$A_P \cdot M$ = Addendum of the pinion.

Where, A_P is a fraction by which the standard addendum of one module for the pinion should be multiplied in order to avoid interference.

It is seen that the addendum of the pinion = $O_1N - O_1P$

$$A_P \cdot M = \frac{MT_P}{2} \sqrt{1 + \frac{T_G}{T_P}\left(\frac{T_G}{T_P} + 2\right) \sin^2 \phi} - \frac{MT_P}{2} \quad \left[\because O_1 P = \gamma = \frac{MT_P}{2}\right]$$

$$= \frac{MT_P}{2}\left[\sqrt{1 + \frac{T_G}{T_P}\left(\frac{T_G}{T_P} + 2\right) \sin^2 \phi} - 1\right]$$

$$A_P = \frac{T_P}{2}\left[\sqrt{1 + \frac{T_G}{T_P}\left(\frac{T_G}{T_P} + 2\right) \sin^2 \phi} - 1\right]$$

$$T_P = \frac{2A_P}{\sqrt{1 + \frac{T_G}{T_P}\left(\frac{T_G}{T_P} + 2\right) \sin^2 \phi} - 1}$$

We know that, gear ratio is given by,

$$G = \text{Gear ratio} = \frac{T_G}{T_P}$$

So equation becomes,

$$T_{P((min))} = \frac{2A_P}{\sqrt{1 + G(G+2)\sin^2 \phi} - 1}$$

This equation gives the minimum number of teeth required on the pinion in order to avoid interference.

Note: If the pinion and gear wheel have equal teeth, then $G = 1$.

Then,

$$T_{P(min)} = \frac{2A_P}{\sqrt{1 + \xi \sin^2 \phi} - 1}$$

Problem

1. A pinion of 24 teeth drives a gear of 60 teeth at a pressure angle of 20°. The pitch radius of the pinion is 38 mm and the outside radius is 41 mm. The pitch radius of the gear is 95 mm and the outside radius is 98.5 mm. Let us calculate the length of action and contact ratio.

Solution:

Given:

$$r_1 = 38mm$$

$$r_2 = 95mm$$

$$r_{a1} = 41mm$$

$$r_{a2} = 98.5mm$$

$$\alpha = 20°$$

$$r_{b1} = r_1 \cos\alpha = 38\cos 20° = 35.7 \text{ mm}$$

$$r_{b2}, r_2 \cos\alpha = 95 \cos 20° = 811.27mm$$

Length of path of contact,

$$L_P = \left(r^2_{a1} - r^2_{b1}\right)^{0.5} + \left(r^2_{a2} - r^2_{b2}\right)^{0.5} - \left(r_1 + r_2\right)\sin\alpha$$

$$L_P = \left(41^2 - 35.7^2\right)^{0.5} + \left(98.5^2 - 811.27^2\right)^{0.5} - \left(38 + 95\right)\sin 20°$$
$$= 20.16 + 41.63 - 45.49$$
$$= 16.30mm$$

Contact ratio,

$$m_c = \frac{L_P}{P_b}$$

$$P_b = 2\pi\frac{r_{b1}}{z_1}$$

$$= 2\pi \times \frac{35.7}{24} = 11.37 \text{ mm}$$

$$m_c = \frac{16.30}{11.34} = 1.744$$

6.3   Contact Ratio of Spur, Helical, Bevel and Worm Gears

Contact ratio is the average number of pairs of teeth in contact. Contact ratio for gears is greater than one.

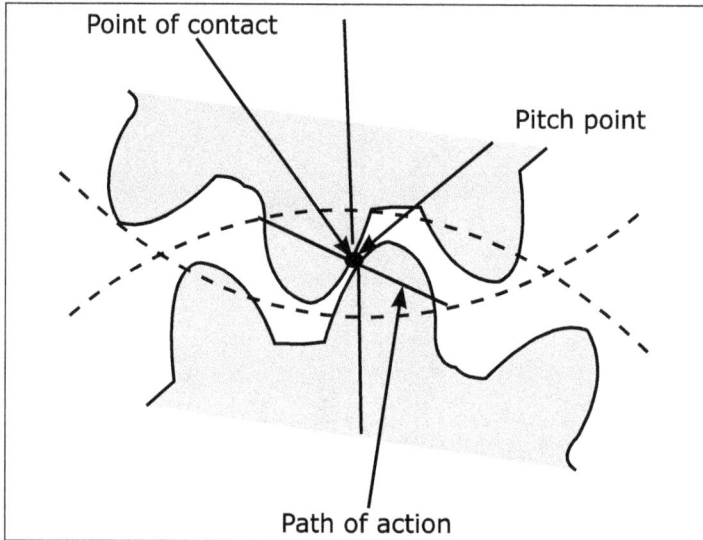

Contact Ration of Spur Gear

During the action of teeth engagement along the path of action in a meshing pair of gear teeth, i.e. from the beginning of contact to the end of contact comprising the length of contact, the load is transmitted by a single tooth of the driving gear for part of the time and by two teeth during rest of the time.

A new pair of teeth comes into action before the preceding pair goes out of action. For continuous contact, the angle of action must be greater than the angle subtended at the centre by the arc representing the circular pitch (called "pitch angle").The relation between these two angles is termed as the 'contact ratio".

The physical significance of the contact ratio lies in the fact that it is a measure of the average number of teeth in contact during the period in which a tooth comes and goes out of contact with the mating gear. Contact begins when the line of action intersects the tip circle of the driven gear. At this point, the flank of driver touches the tip of the driven gear.

Contact ends when the line of action intersects the tip circle of the driver. At this point, the tip of the driver just leaves the flank of the driven gear. These two points are shown as A and B in the below figure and the portion of the line of contact AB of T_1T_2 is called the length of contact or the length of action.

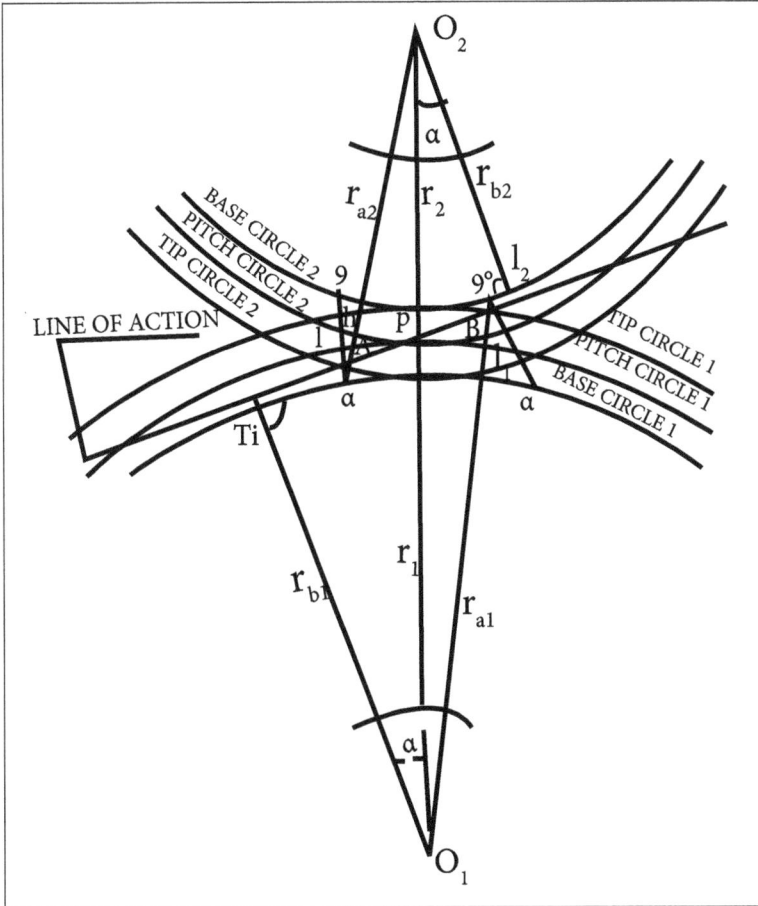

Derivation of Contact Ratio.

From the above figure,

Contact ratio (CR) is given by,

$$CR = \frac{\text{Angle of action}}{\text{Pitch angle}} = \varepsilon \, (\text{IS symbol})$$

$$\text{Contact ratio} \, (CR) = \frac{\text{Angle of action}}{\text{Pitch angle}} = \frac{\text{Arc of action}}{\text{Circular pitch}} = \frac{\text{Length of action}}{\text{Base pitch}}$$

$$\text{Therefore}, CR = \frac{\sqrt{r_{a_1}^2 - r_{b_1}^2} + \sqrt{r_{a_2}^2 - r_{b_2}^2} - a \sin \alpha}{p \cos \alpha}$$

$$= \frac{\sqrt{r_{a_1}^2 - r_{b_1}^2} + \sqrt{r_{a_2}^2 - r_{b_2}^2} - a \sin \alpha}{p_b}$$

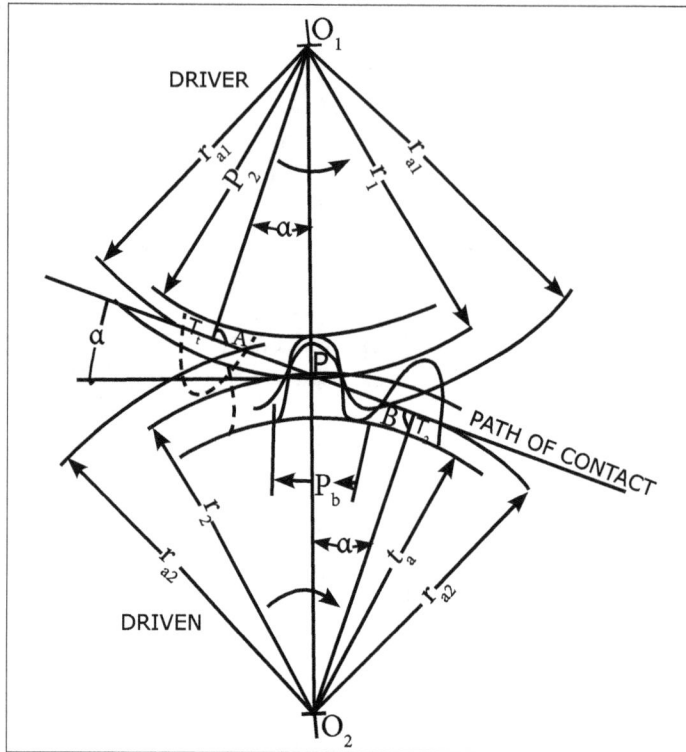

Contact Ratio.

In the figure,

Beginning of contact at "A" (Shown in dotted lines).

End of contact at "B" (Shown in solid lines).

From the figure,

$$\text{Contact Ratio} = \frac{\text{Length of Contact}}{\text{Base Pitch}} = \frac{AB}{P_b}$$

For a gear mating with a rack, the contact ratio is given by,

$$CR = \frac{\sqrt{r_a^2 - r_b^2} - r \sin \alpha + \dfrac{h_r}{\sin \alpha}}{p \cos \alpha}$$

Where, r_a, r_b and r relate to the gear and h_r is the addendum of the rack and is usually equal to the module m.

6.3.1 Helical Gears

The teeth on helical gears are cut at an angle to the face of the gear. When two teeth on a helical gear system engage, the contact starts at one end of the tooth and gradually spreads as the gears rotate, until the two teeth are in full engagement.

This gradual engagement makes the helical gears operate much more smoothly and quietly than spur gears. For this reason, helical gears are used in almost all car transmissions.

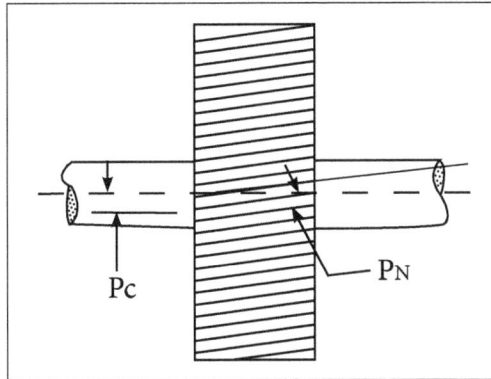

Helical gear.

Because of the angle of teeth on helical gears, they create a thrust load on the gear when they mesh. Devices that use helical gears have bearings that can support this thrust load.

One interesting thing about helical gears is that if the angles of the gear teeth are correct, they can be mounted on perpendicular shafts, adjusting the rotation angle by 90 degrees.

The use of helical gears is most common in automobiles, turbines and high speed applications. It can be seen that the teeth of the two wheels are of opposite hand. The helixes may be right handed on one wheel and left handed on the other:

- Normal pitch: It is the distance between similar faces of adjacent teeth, along a helix on the pitch cylinder normal to the teeth. It is denoted by p_N.

- Axial pitch: It is the distance measured parallel to the axis, between similar faces of adjacent teeth. It is same as circular pitch and is therefore denoted by p_c. If α is the helix angle, then circular pitch,

$$P_c = \frac{P_N}{\cos \alpha}$$

Advantages of Helical Gear

There are three main reasons why helical gears are preferred than spur gears. They are:

- Noise: Helical gears produce less noise than spur gears of equivalent quality because the total contact ratio is increased.

- Load carrying capacity: Helical gears have a greater load carrying capacity than equivalent size spur gears because the total length of the line of contact is increased.

- Manufacturing: A limited number of standard cutters are used to cut a wide variety of helical gears simply by varying the helix angle.

Disadvantage of Helical Gear

Since the helical gears are inclined to the axis of rotation it is subjected to axial thrust loads. This problem can be eliminated by using double helical gears (Herringbone gears):

- Helix angle: It is constant angle made by the helices with the axis of rotation.

- Axial pitch: It is the distance parallel to the axis between similar faces of adjacent teeth. It is same as circular pitch and is therefore denoted by P_c.

- Normal pitch: It is the distance between similar faces of adjacent teeth along a helix on the pitch cylinders normal to the teeth. It is denoted by P_N.

$$P_N = P_c \cos \beta,$$

$$\tan \alpha_N = \tan\alpha \cos\beta.$$

Where,

α_N = Normal pressure angle,

α = Pressure angle.

Face width: In order to have more than one pair of teeth in contact, the tooth displacement (i.e., the advancement of one end of tooth over the other end) or overlap ought to be at least equal to the axial pitch such that, overlap,

$$P_c = b \tan \beta \, p_c$$

Formative or Equivalent Number of Teeth for Helical Gear

The formative or equivalent number of teeth for a helical gear may be defined as the number of teeth that will be generated on the surface of a cylinder having a radius equal to the radius of curvature at a point at the tip of the minor axis of an ellipse obtained by taking a section of the gear in the normal plane. Mathematically, formative or equivalent number of teeth on a helical gear,

$$Z_E = Z/\cos^3.\beta$$

Where,

Z = Actual number of teeth on a helical gear,

β = Helix angle.

Proportion of Helical Gears

Pressure angle in the plane of rotation α = 15° to 25°

Helix angle, β = 20° - 45°

Addendum = 0.8 m (maximum)

Dedendum = 1.0 m

Minimum total depth = 1.8 m (maximum)

Minimum clearance = 0.2 m

Thickness of tooth = 1.5708 m

Design

In helical gears, the contact between mating teeth is gradual, starting at one end and moving along the teeth so that at any instant the line of contact runs diagonally across the teeth. Therefore, in order to find the strength of helical gear, a modified lewis equation is used.

It is given by, $F_t = \sigma_o C_v b\pi my'$

Where,

F_t, σ_o, C_v, b, π, m, as usual, with same meanings.

And y' = Tooth from factor or lewis factor corresponding to the formative or virtual or equivalent number of teeth.

Item	Equation
For low-angel helical gears when v is less than 5 m/s	$C_v \dfrac{4.58}{4.58 + v}$
For all helical and herringbone gears when v is 5 to 10 m/s	$C_v \dfrac{6.1}{6.1 + v}$
For gears when v is 10 to 20 m/s (Barth's formula)	$C_v \dfrac{15.25}{15.25 + v}$
For precision gear with v greater than 20 m/s	$C_v \dfrac{5.55}{5.55 + \sqrt{v}}$
For nonmetallic gears	$C_v \dfrac{0.7625}{1.0167 + v} + 0.25$

- The dynamic tooth load, $F_d = F_t + F_i$

Where,

$$F_i = \frac{K_3 v\left(cb\ cos^2\beta + F_t\right)\cos\beta}{K_3 v + \left(cb\ cos^2\beta + F_t\right)^{1/2}}$$

K_s = 20.67 in SI units

= 6.60 in metric units.

- The static tooth load or endurance strength of the tooth is given by,

$F_s = \sigma_e b\pi\ my' \geq F_d$

The maximum or limiting wear tooth load for helical gears is given by,

$$F_w = \frac{d_1\ bQ\ K}{cos^2\beta} \geq F_d$$

In this case,

Where, K = the load stress factor.

$$K = \frac{\left(\sigma_{es}\right)^2 \sin\alpha_N}{1.4}\left[\frac{1}{E_1} + \frac{1}{E_2}\right]$$

Wear Loads of the Helical Gear

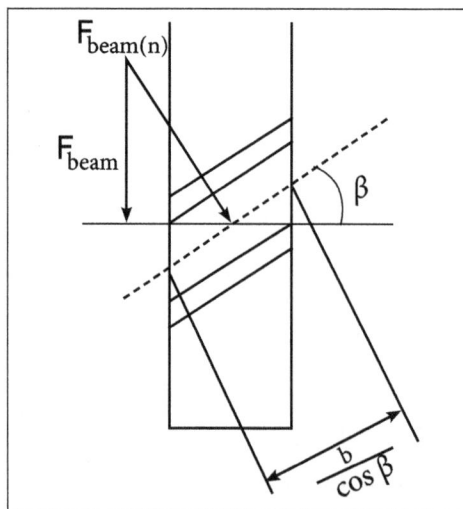

Wear loads of the helical gear.

The wear strength equation of the helical gear is obtained by modifying the corresponding equation of the spur gear. For this purpose, an equivalent formative pinion and gear is considered. The wear equation of the spur gear is given as:

$$F_{wear} = d_1 \, b \, Q \, K,$$

$$F_{wear(n)} = \frac{d_1 \, b \, Q \, K}{\cos^3 \beta},$$

F_{wear} with $F_{wear\,(n)}$, i.e., Wear strength normal to the tooth.

b with b/cos β, i.e. Normal face width,

d_1 with $d_1/\cos^2\beta$, i.e. Pitch circle diameter of the formative pinion.

Contact Ration of Helical Gear

The derivation of the contact ratio and its implications has been dealt with in detail. The contact ratio of a pair of helical gears in mesh can be found in a similar way. That due to the effect of the face advance in a helical gear, an extra amount of contact ratio is created.

This face advance is due to the helical orientation of the tooth along the length of the tooth, covering the width of the gear. The contact ratio due to face advance is called the face contact ratio and is given by,

$$CR_{FA} = \frac{\text{Face advance}}{\text{Transverse circular pitch}} = \frac{b \tan \beta}{p_t}$$

Now,

$$p_t = \frac{\pi m}{\cos \beta} \qquad \therefore CR_{FA} = \frac{b \tan \beta}{\pi m} \cos \beta = \frac{b \sin \beta}{\pi m}$$

The face contact ratio is also known as the axial contact ratio and the overlap ratio. Due to this, the total contact ratio in case of a helical gear is greater than that of a spur gear.

The transverse contact ratio in case of a pair of helical gears in mesh can be found in a similar manner as in the case of spur gears. It is given by,

$$CR_T = \frac{\sqrt{r_{a_1}^2 - r_{b_1}^2} + \sqrt{r_{a2}^2 - r_{b_2}^2} - a \sin \alpha_t}{p_{bt}}$$

For corrected helical gears, the corrected values of r_{a1}, r_{a2} a and the centre distance a are to be inserted. Moreover, α_t will be replaced by the working pressure angle in the transverse section α_{tw}.

The total contact ratio (or, simply, the contact ratio) of the helical gearing is a summation of the above two contact ratios. Therefore,

$$CR = CR_{FA} + CR_T$$

$$= \frac{b\tan\beta}{p_t} + \frac{\sqrt{r_{a_1}^2 - r_{b_1}^2} + \sqrt{r_{a2}^2 - r_{b_2}^2} - a\sin\alpha_t}{p_{bt}} \qquad ...(i)$$

Since the circular pitch and the base pitch in the transverse section are related by the expression.

Transverse base pitch p_{bt} = Transverse circular pitch p_t x $\cos\alpha_t$

Equation (i),can be written as,

$$CR = \frac{b\sin\beta\cos\alpha_t + \cos\beta\left[\sqrt{r_{a_1}^2 - r_{b_1}^2} + \sqrt{r_{a2}^2 - r_{b_2}^2} - a\sin\alpha_t\right]}{\pi m \cos\alpha_t}$$

6.3.2 Bevel Gear

Bevel gears transmit power between two intersecting shafts at any angle or between non- intersecting shafts. They are classified as straight and spiral tooth bevel and hypoid gears. Bevel gears are used to transmit power between two intersecting shaft and bevel gears are commonly used in automotive differentials.

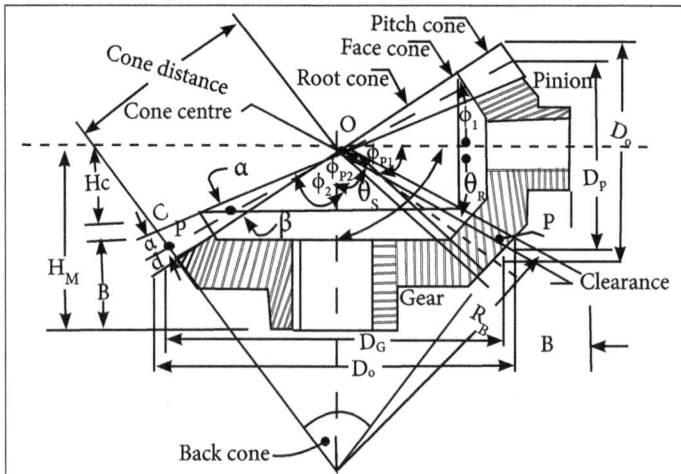

Bevel gear.

When intersecting shafts are connected by gears, the pitch cones (analogous to the pitch cylinders of spur and helical gears) are tangent along an element, with their apexes at the intersection of the shafts as shown where two bevel gears are in mesh. The size and shape of the teeth are defined at the large end, where they intersect the back cones.

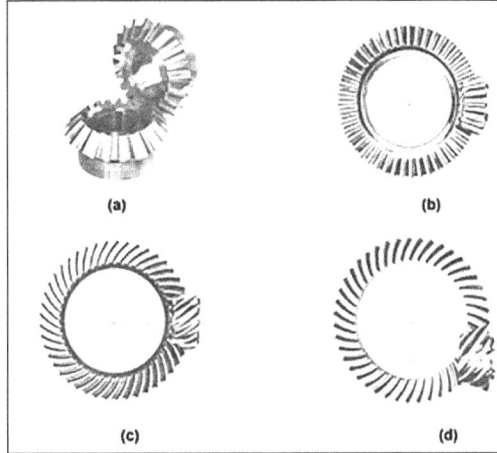

(a) Bevel gear, (b) Straight bevel gear, (c) Spiral bevel gear and (d) Hypoid gear.

Pitch cone and back cone elements are perpendicular to each other. The tooth profiles resemble those of spur gears having pitch radii equal to the developed back cone radii. The gear ratio can be determined from the number of teeth, the pitch diameters or the pitch cone angles as,

$$i = \frac{\omega_1}{\omega_2} = \frac{n_1}{n_2} = \frac{z_2}{z_1} = \frac{d_2}{d_1} = \tan\gamma_2 = \cot\gamma_1$$

Tooth Bending Stress

The equation for bevel gear bending stress is the same as for spur gears as shown below:

$$\sigma_b = \frac{F_t}{bmJ} K_v K_o K_m$$

The following terms in connection with bevel gears are important from the subject point of view:

- Pitch cone: A cone containing the pitch elements of the teeth.

- Cone centre: It is the apex of the pitch cone. It may be defined as the point where the axes of two mating gears intersect each other.

- Pitch angle: It is the angle made by the pitch line with the axis of the shaft. It is denoted by 'θ_p'.

- Cone distance: It is the length of the pitch cone element. It is also called as a pitch cone radius. It is denoted by 'OP'. Mathematically, cone distance or pitch cone radius,

$$OP = \frac{\text{Pitch radius}}{\sin\theta_p} = \frac{D_p/2}{\sin\theta_{p1}} = \frac{D_G/2}{\sin\theta_{p2}}$$

- Addendum angle: The angle subtended by the addendum of the tooth at the cone centre. It is denoted by 'α'. Mathematically, addendum angle,

$$\alpha = \tan^{-1}\left(\frac{a}{OP}\right)$$

Where,

   a = Addendum.

   OP = Cone distance.

- Dedendum angle: It is the angle subtended by the dedendum of the tooth at the cone centre. It is denoted by 'β'. Mathematically, dedendum angle,

$$\beta = \tan^{-1}\left(\frac{d}{OP}\right)$$

Where,

   d = Dedendum.

   OP = Cone distance.

- Face angle: It is the angle subtended by the face of the tooth at the cone centre. It is denoted by 'φ'. The face angle is equal to the pitch angle plus addendum angle.

- Root angle: It is the angle subtended by the root of the tooth at the cone centre. It is denoted by 'θ_R'. It is equal to the pitch angle minus dedendum angle.

- Back or normal cone: It is an imaginary cone, which is perpendicular to the pitch cone at the end of the tooth.

- Back cone distance: It is the length of the back cone. It is denoted by 'R_B'. It is also called as back cone radius.

- Backing: The distance of the pitch point (P) from the back of the boss, parallel to the pitch point of the gear. It is denoted by 'B'.

- Crown height: It is the distance of the Crown Point (C) from the cone centre (O), parallel to the axis of the gear. It is denoted by 'H_C'.

- Mounting height: It is the distance of the back of the boss from the cone centre. It is denoted by 'HM'.

- Pitch diameter: It is the diameter of the largest pitch circle.

- Outside or addendum cone diameter: It is the maximum diameter of the teeth of the gear. It is equal to the diameter of the blank from which the gear can be cut. This is the outside Diameter,

$$D_o = D_p + 2a \, \text{Cos} \, \theta_p$$

Where,

D_p = Pitch circle diameter.

a = Addendum.

θ_p = Pitch angle.

- Inside or dedendum cone diameter: The inside or the dedendum cone diameter is given by,

$$D_d = D_p - 2d \, \text{Cos} \, \theta_p$$

Where,

D_d = Inside diameter.

D = Dedendum.

Strength of Bevel Gears

The strength of a bevel gear tooth is the modified form of the Lewis equation for the tangential tooth load is given as follows:

$$W_T = (\sigma_o \times C_V) \, b.\pi m.y' \left(\frac{L-b}{L} \right)$$

Where,

σ_o = Allowable static stress,

L = Slant height of pitch cone,

C_v = Velocity factor,

$= \dfrac{3}{3+v}$, for teeth cut by form cutters,

$= \dfrac{6}{6+v}$, for teeth generated with precision machines,

b = Face width,

m = Module,

v = Peripheral speed in m/s,

y' = Tooth form factor (or Lewis factor) for the equivalent number of teeth,

$$= \sqrt{\left(\frac{D_G}{2}\right)^2 + \left(\frac{D_P}{2}\right)^2}$$

D_G = Pitch diameter of the gear,

D_P = Pitch diameter of the pinion.

Design of a Shaft for Bevel Gears

- Determine the torque acting on the pinion. It is given by,

$$T = \frac{P \times 60}{2\pi N_P} N-m$$

P = Power transmitted in watts.

N_P = Speed of the pinion in rpm.

- Determine the tangential force (W_T) acting at the mean radius (R_m) of the pinion. We know that,

$$W_T = T / R_m.$$

- Determine the axial and radial forces acting on the pinion shaft.
- Determine the resultant bending moment on the pinion shaft as follows:

The bending moment due to W_{RH} and W_{RV} is given by,

$$M_1 = W_{RV} \times \text{Overhang} - W_{RH} \times R_m,$$

And bending moment due to W_T,

$$M_2 = W_T \times \text{Overhang}.$$

Resultant bending moment,

$$M = \sqrt{(M_1)^2 + (M_2)^2}$$

- Since, the shaft is subjected to twisting moment (T) and resultant bending moment (M). Hence, equivalent twisting moment,

$$T_\theta = \sqrt{M^2 + T^2}$$

- The diameter of the pinion shaft may be obtained by using the torsion equation. We know that,

$$T_\theta = \frac{\pi}{16} \times \tau (d_p)^3$$

Where,

d_p = Diameter of the pinion shaft.

τ = Shear stress for the material of the pinion shaft.

The same procedure may be adopted to find the diameter of the gear shaft.

Proportions for Bevel Gear

The proportions for the bevel gears may be taken as follows:

- Addendum, a = 1 m.
- Dedendum, d = 1.2 m.
- Clearance = 0.2 m.
- Working depth = 2 m.
- Thickness of tooth = 1.5708 m.

Where, m is the module.

6.3.3 Worm Gear

Worm gears are used when large gear reductions are needed. It is common for worm gears to have reductions of 20:1 and even up to 300:1 or greater.

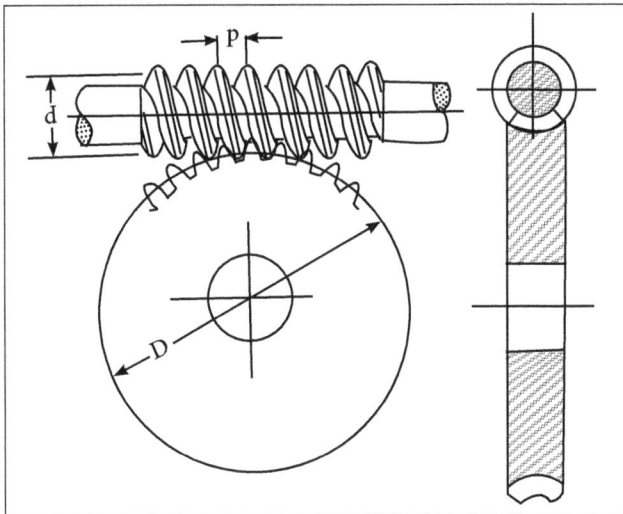

Worm gear.

Many worm gears have an interesting property that no other gear set has, the worm can easily turn the gear, but the gear cannot turn the worm. This is because the angle on the worm is so shallow that when the gear tries to spin it, the friction between the gear and the worm holds the worm in place.

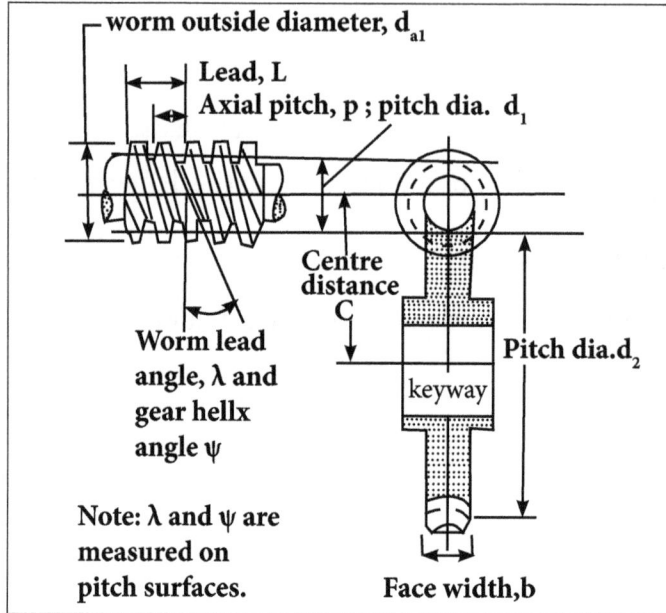

Nomenclature of worm gear.

This feature is useful for machines such as conveyor systems, in which the locking feature can act as a brake for the conveyor when the motor is not turning. One other very interesting usage of worm gears is in the Torsen differential, which is used on some high-performance cars and trucks.

Worm Gear Terminology

The geometry of a worm is similar to that of a power screw. Rotation of the worm simulates a linearly advancing involute rack. The geometry of a worm gear is similar to that of a helical gear, except that the teeth are curved to envelop the worm. Enveloping the gear gives a greater area of contact but requires extremely precise mounting.

- As with a spur or helical gear, the pitch diameter of a worm gear is related to its circular pitch and number of teeth Z by the formula,

$$d_2 = \frac{Z_2 p}{\pi}$$

- When the angle is 90° between the non-intersecting shafts, the worm lead angle λ is equal to the gear helix angle Ψ. Angles λ and Ψ have the same hand.

- The pitch diameter of a worm is not a function of its number of threads, Z_1.

- This means that the velocity ratio of a worm gear set is determined by the ratio of gear teeth to worm threads. It is not equal to the ratio of gear and worm diameters.

$$\frac{\omega_1}{\omega_2} = \frac{Z_2}{Z_1}$$

- Worm gears usually have at least 24 teeth and the number of gear teeth plus worm threads should be more than 40.

$$Z_1 + Z_2 > 40$$

- A worm of any pitch diameter can be made with any number of threads and any axial pitch.

- Integral worms cut directly on the shaft cam, of course, have a smaller diameter than that of shell worms, which are made separately.

- Shell worms are bored to slip over the shaft and are driven by splines, key or pin.

- Strength considerations seldom permit a shell worm to have a pitch diameter less than

$$d_1 = 2.4p + 1.1$$

- The face width of the gear should not exceed half the worm outside diameter.

$$b \leq 0.5d_{a1}$$

Advantages of Worm Gear Drive

- The worm gear drives can be used for speed ratio as high as 300:1.

- The operation is smooth and silent.

- The worm gear drives are compact compared with equivalent spur or helical gears for the same speed reduction.

- The worm gear drives are irreversible. It means that the motion cannot be transmitted from worm wheel to the work. This property of irreversible is advantages in load hoisting application like cranes and lifts.

6.3.4 Interference in Involute Gears

The figure below shows a pinion with centre O_1, in mesh with wheel or gear with centre O_2. MN is the common tangent to the base circles and KL is the path of contact between the two mating teeth. If the radius of the addendum circle of pinion is increased to O_1N, the point L will move to N.

When this radius is further increased, the point of contact L will be on the inside of base circle and not on the involute profile of tooth on wheel. The tip of tooth on the pinion will then undercut the tooth on the wheel at the root and remove part of the involute profile of tooth on the wheel. This effect is known as interference.

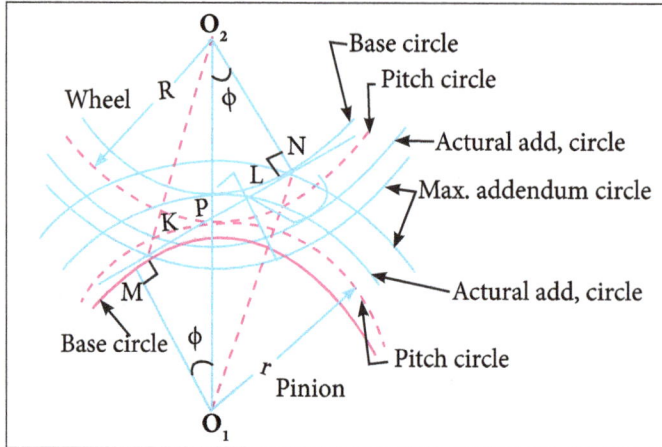

Interference in involute gears.

If the radius of the addendum circle of the wheel increases beyond O_2M, then the tip of tooth on wheel will cause interference with the tooth on pinion. The points M and N are called interference points.

When interference is just avoided, the maximum length of path of contact is MN when the maximum addendum circles for pinion and wheel pass through the points of tangency N and M.

Maximum length of path of approach, $MP = r \sin \phi$.

Maximum length of path of recess $PN = \sin \phi$.

Therefore, maximum length of path of contact.

$MN = MP + PN = r \sin \phi + R \sin \phi = (r + R) \sin \phi$.

And, maximum length of arc of contact.

$$= \frac{(r+R)\sin\phi}{\cos\phi} = (r+R)\tan\phi$$

Undercutting of Teeth

To ignore interference, a portion of the radial part of the tooth may be cut out. This undercutting shall provide clearance among the gear teeth having the pinion teeth. But the undercutting makes teeth weaker at the root. Thus, such a method may be utilized only for those gears that run at very low speed. Shown in figure illustrates the portion to be cut for undercutting.

To decrease space needs below pitch circle of pinion and above that base circle, the stub teeth may be utilized for mating gear. In stub teeth the addendum of gear is given as '0.6 m' instead of standard addendum of '1m'.

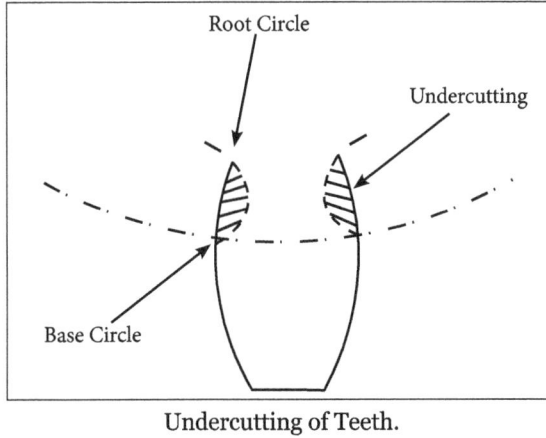

Undercutting of Teeth.

The phenomenon when the tip of tooth undercuts the root on its mating gear is known as interference.

In order to avoid interference the following methods may be employed:

- The height of the teeth may be reduced,
- The pressure angle may be increased,
- The radial flank of the pinion may be cut back,
- The face of the gear tooth may be relieved.

Involute Teeth

An involute of a circle is a plane curve generated by a point on a tangent, which rolls on the circle without slipping or by a point on a taut string which in unwrapped from a reel as shown in figure (a). In connection with toothed wheels, the circle is known as base circle. The involute is traced as follows:

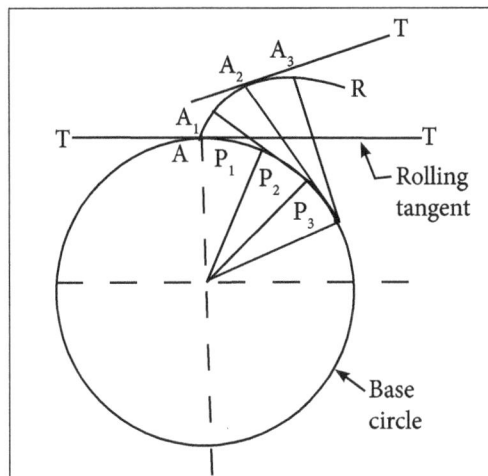

(a) Construction of involute.

Let A be the starting point of the involute. The base circle is divided into equal number of parts e.g. AP_1, P_1P_2, P_2P_3 etc. The tangents at P_1, P_2, P_3 etc. Are drawn and the length P_1A_1, P_2A_2, P_3A_3 equal to the arc AP_1, AP_2 and AP_3 are set off. Joining the points A, A_1, A_2, A_3 etc. we obtain the involute curve AR. A little consideration will show that at any instant A_3, the tangent A_3T to the involute is perpendicular to P_3A_3 and P_3A_3 is the normal to the involute. In other words, normal at any point of an involute is a tangent to the circle.

Now O_1 and O_2 be the fixed centres of the two base circles as shown in figure (b). Let the corresponding involutes AB and A_1B_1 be in contact at point Q. MQ and NQ are normals to the involutes at Q and Q' are tangents to base circles. Since the normal of an involute at a given point is the tangent drawn from that point to the base circle, therefore the common normal MN at Q is also the common tangent to the two base circles.

The common normal MN intersects the line of centres O_1O_2 at the fixed point P. Therefore the involute teeth satisfy the fundamental condition of constant velocity ratio.

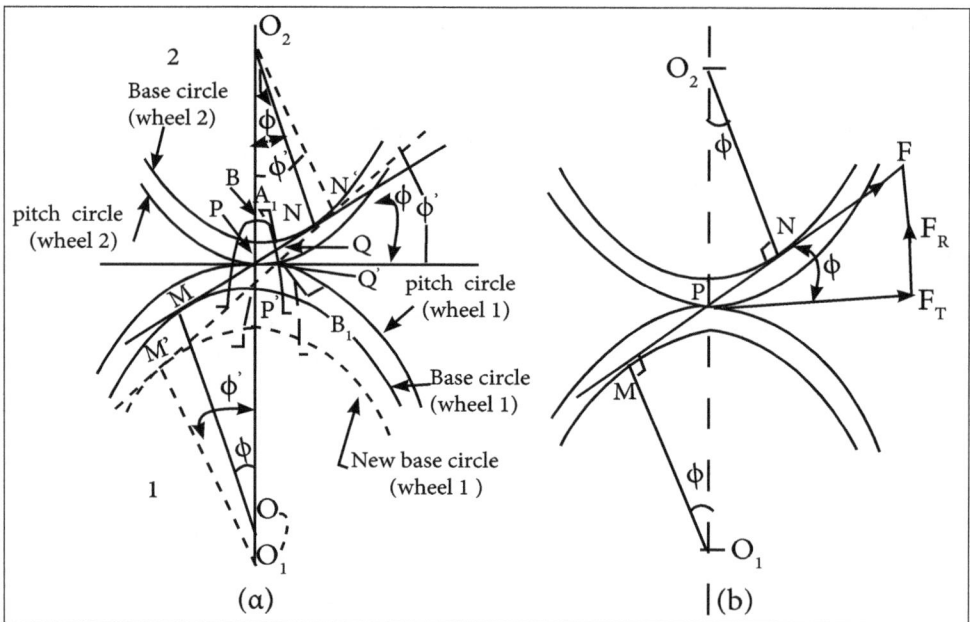

(b) Involute teeth.

From similar triangles O_2NP and O_1MP,

$$\frac{O_1M}{O_2N} = \frac{O_1P}{O_2P} = \frac{\omega_2}{\omega_1} \qquad ...(1)$$

Which determines the ratio of the radii of the two base circles. The radii of the base circles is given by,

$$O_1M = O_1P\cos \varphi \text{ and } O_2P \cos \varphi$$

Also the centre distance between the base circles,

$$O_1O_2 = O_1P + O_2P = \frac{O_1M}{\cos\phi} + \frac{O_2N}{\cos\phi} = \frac{O_1M + O_2N}{\cos\phi}$$

Where, ϕ is the pressure angle or the angle of obliquity. It is the angle which is the common normal to the base circles. The common tangent to the pitch circles.

The power is being transmitted; the high tooth pressure is exerted along the common normal through the pitch point. This force may be resolved into tangential and radial or normal components. These components act along and at right angles to the common tangent to the pitch circles

If F is the maximum tooth pressure as shown in figure (b). then,

Tangential force, $F_T = F \cos\phi$

Radial or normal force, $F_R = F \sin\phi$

$\therefore$ Torque exerted on the gear shaft $= F_T \times r$

Where,

      r = Pitch circle radius of the gear.

Problems

1. A pair 20° full depth in-volute gears having 30 and 50 teeth respectively of module 4 mm are in mesh, the smaller gear rotates at 1000 r.p.m. Let us determine: (i) sliding velocities at engagement and at disengagement of a pair of teeth and (ii) contact ratio.

Solution:

Given:

Pressure angle, $\phi = 20°$

No. of teeth on smaller gear, t = 30

No. of teeth on large gear, T = 50

Gear ratio,

$$G = \frac{T}{t} = \frac{50}{30}$$

Module, m = 4 mm

Speed of smaller gear, $N_1 = 1000$ rpm

$\therefore$ Angular speed of smaller gear,

$$\omega_1 = \frac{2\pi N_1}{60} = \frac{2\pi \times 1000}{60}$$

$$= 104.67 \text{ rad/s.}$$

To find:

- Sliding velocities at engagement and at disengagement of a pair of teeth.

- Contact ratio.

(i) Sliding velocities at engagement and at disengagement:

First calculate the length of path of approach and length of path of recess. For these, we require the values of R_o, R, r_o and r. Also we require the values of addendum of the wheel and addendum of the pinion.

Now addendum of wheel (or of larger gear) is given by equation as:

Addendum of wheel,

$$= \frac{m \times T}{2} \left[\sqrt{\left\{ 1 + \frac{1}{G} \left(\frac{1}{G} + 2 \right) \sin^2 \phi \right\}} - 1 \right]$$

$$= \frac{4 \times 50}{2} \left[\sqrt{\left\{ 1 + \frac{30}{50} \left(\frac{30}{50} + 2 \right) \sin^2 20^\circ \right\}} - 1 \right] \text{mm} \quad \left(\because G = \frac{T}{t} = \frac{50}{30} \therefore \frac{1}{G} = \frac{30}{50} \right)$$

$$= 100 \left[\sqrt{\left\{ 1 + 0.6 \times 2.6 \times 0.342^2 \right\}} - 1 \right]$$

$$= 100 \left[\sqrt{1.1824} - 1 \right] = 9 \text{mm}$$

Also the addendum of the pinion (or of smaller gear) is given by equation as,

Addendum of pinion,

$$= \frac{m \times t}{2} \left[\sqrt{\left\{ 1 + \frac{T}{t} \left(\frac{T}{t} + 2 \right) \sin^2 \phi \right\}} - 1 \right]$$

$$= \frac{4 \times 30}{2} \left[\sqrt{\left\{ 1 + \frac{5}{30} \left(\frac{50}{30} + 2 \right) \sin^2 20^\circ \right\}} - 1 \right]$$

$$=60 \left[\left[\sqrt{1+1.667\,(3.667)\times 0.342^2} \right] -1 \right]$$

$$=60 \left[\sqrt{1+0.7154} -1 \right] =60(1.309-1)\ \text{mm}$$

$$= 18.54\ \text{mm}$$

Now pitch circle radius of wheel,

$$R=\frac{m\times T}{2} = \frac{4\times 50}{2}=100\,\text{mm}$$

∴ Radius of addendum circle of wheel,

$$R_0 = R + \text{addendum of wheel} = 100 + 9 = 109\ \text{mm}$$

Now, pitch radius of pinion,

$$r=\frac{m\times t}{2} = \frac{4\times 30}{2}=60\,\text{mm}$$

∴ Radius of addendum circle of pinion,

$$r_0 = r + \text{addendum of pinion} = 60 + 18.54 = 78.54\ \text{mm}.$$

Now length of the path of approach (i.e., the path of contact when engagement occurs) from equation,

$$= \sqrt{R_0^2 - R^2\,\cos^2\phi} - R\,\sin\phi$$

$$= \sqrt{109^2 - 100^2\,\cos^2 20^\circ} - 100\times\sin 20^\circ$$

$$= \sqrt{11881 - 8830} - 34.255$$

$$= 21.03\ \text{mm}$$

Length of the path of recess (i.e., the path of contact when disengagement occurs) [Refer to equation],

$$= \sqrt{r_0^2 - r^2\,\cos^2\phi} - r\,\sin\phi$$

$$= \sqrt{78.54^2 - 60^2\,\cos^2 20^\circ} - 60\,\sin 20^\circ$$

$$= \sqrt{6168.53 - 3174.19} - 20.52 = 54.72 - 20.52$$

$$= 34.2\ \text{mm}$$

Now, let us find the sliding velocities at engagement and disengagement of a pair of teeth.

Let,

ω_2 = Angular speed of the wheel (or large gear).

We know from equation that,

$$\frac{\omega_2}{\omega_1} = \frac{t}{T}$$

$$\therefore \omega_2 = \omega_1 \times \frac{t}{T} = 104.67 \times \frac{30}{50} \text{ rad/s} = 62.8 \text{ rad/s}$$

$\therefore$ Sliding velocity at engagement of a pair of teeth,

= $(\omega_1 + \omega_2) \times$ length of path of approach

= $(104.67 + 62.8) \times 21.03$

= 3521.9 mm/s

= 3.522 m/s

And sliding velocity at disengagement position,

= $(\omega_1 + \omega_2) \times$ length of path of recess

= $(104.67 + 62.8) \times 34.2$

= 5727.5 mm/s = 5.727 m/s

(ii) Contact ratio:

$$\text{Contact ratio} = \frac{\text{Length of arc of contact}}{\text{Circular pitch}}$$

But length of arc of contact,

$$= \frac{\text{Length of path of contact}}{\cos\phi}$$

$$= \frac{\text{Length of path of approach} + \text{Length of path of recess}}{\cos\phi}$$

$$= \frac{21.03 + 34.2}{\cos 20^\circ} = 58.8 \text{ mm}$$

Also we know that,

Circular pitch = π × m = π × 4 = 12.568 mm

∴ substituting the above values in equation (r). We get,

$$= \frac{58.8}{12.568} = 4.67 \, \text{say } 5$$

2. A pair of helical gears is to transmit 15 kW. The teeth are 20° stub in diametral plane and have a helix angle of 45°. The pinion runs at 10 000 rpm and has 80 mm pitch diameter. The gear has 320 mm pitch diameter. If the gears are made of cast steel having allowable static strength of 100 MPa. Let us determine a suitable module and face width from static strength considerations and check the gears for wear, given σ_{es} = 618 MPa.

Solution:

Given:

P = 15 kW = 15 × 10³ W,

φ = 20°; α = 45°,

NP = 10 000 rpm,

D_P = 80 mm = 0.08 m,

D_G = 320 mm = 0.32 m,

$\sigma_{OP} = \sigma_{OG}$ = 100 MPa = 100 N/mm2,

σ_{es} = 618 MPa = 618 N/mm2.

Let,

m = Module in mm,

b = Face width in mm.

Since, both the pinion and gear are made of the same material (i.e. cast steel), therefore the pinion is weaker. Thus the design will be based upon the pinion. We know that the torque transmitted by the pinion,

$$T = \frac{P \times 60}{2\pi N_P} = \frac{15 \times 10^3 \times 60}{2\pi \times 10000} = 14.32 \, \text{N-m}$$

Tangential tooth load on the pinion,

$$W_T = \frac{T}{D_P / 2} = \frac{14.32}{0.08 / 2} = 358 \, \text{N}$$

We know that number of teeth on the pinion,

$$T_p = D_p/m = 80/m$$

And formative or equivalent number of teeth for the pinion,

$$T_E = \frac{T_p}{\cos^3 \alpha} = \frac{80/m}{\cos^3 45°} = \frac{80/m}{(0.707)^3} = \frac{226.4}{m}$$

∴ Tooth form factor for the pinion for 20° stub teeth,

$$y'_p = 0.175 - \frac{0.841}{T_E} = 0.175 - \frac{0.841}{226.4/m} = 0.175 - 0.0037 \, m$$

We know that peripheral velocity,

$$v = \frac{\pi D_p . N_P}{60} = \frac{\pi \times 0.08 \times 10000}{60} = 42 \, m/s$$

∴ Velocity factor,

$$C_v = \frac{0.75}{0.75 + \sqrt{v}} = \frac{0.75}{0.75 + \sqrt{42}} = 0.104$$

Since, the maximum face width (b) for helical gears may be taken as 12.5 m to 20 m, where m is the module, therefore let us take,

$$b = 12.5m$$

We know that the tangential tooth load (W$_T$),

$$358 = (\sigma_{op}.C_v) \, b.\pi \, m.y'_p$$

$$= (100 \times 0.104) \, 12.5m \times \pi m \, (0.175 - 0.0037)$$

$$= 409m^2(0.175 - 0.0037m) = 72m^2 - 1.5m^3$$

Solving this expression by hit and trial method, we find that,

$$m = 2.3 \, say \, 2.5 \, mm$$

$$b = 12.5m = 12.5 \times 2.5$$

$$= 31.25 say \, 32mm.$$

3. A helical cast steel gear with 30° helix angle has to transmit 35 kW at 1500 rpm. If the gear has 24 teeth, let us determine the necessary module, pitch diameter and face width for 20° full depth teeth. The static stress for cast steel may be taken as 56 MPa.

The width of face may be taken as 3 times the normal pitch. Let us also the end thrust on the gear. The tooth factor for 20° full depth involute gear may be taken as 0.154-(0.912/T_E)where T_E represents the equivalent number of teeth.

Solution:

Given:

$\alpha = 30°$;

$P = 35\ kW = 35 \times 103\ W$;

$N = 1500\ rpm$;

$T_G = 24$;

$\varphi = 20°$;

$56\ MPa = 56\ N/mm^2$;

$b = 3 \times$ Normal pitch $= 3P_N$

Module,

Let m = Module in mm,

D_G = Pitch circle diameter of the gear in mm.

We know that torque transmitted by the gear,

$$T\frac{P \times 60}{2\pi N} = \frac{35 \times 10^3 \times 60}{2\pi \times 1500} = 223\ N-m = 223 \times 10^3\ N-mm$$

Formative or equivalent number of teeth,

$$T_E = \frac{T_G}{\cos^3 \alpha} = \frac{24}{\cos^3 30°} = \frac{24}{(0.866)^3} = 37$$

Tooth factor,

$$y' = 0.154 - \frac{0.912}{T_E} = 0.154 - \frac{0.912}{37} = 0.129$$

We know that the tangential tooth load,

$$W_T = \frac{T}{D_G/2} = \frac{2T}{D_G} = \frac{2T}{m \times T_G} \qquad ...(\because D_G = m.T_G)$$

$$= \frac{2 \times 223 \times 10^3}{m \times 24} = \frac{18600}{m}N$$

Peripheral velocity,

$$v = \frac{\pi D_G \cdot N}{60} = \frac{\pi \cdot m \cdot T_G \cdot N}{60} \, mm/s \qquad \qquad ...(D_G \text{ and m are in mm})$$

$$= \frac{\pi \times m \times 24 \times 1500}{60} = 1.885 \, mm/s$$

Let us take velocity factor,

$$C_v = \frac{15}{15+v} = \frac{15}{15+1.885 \, m}$$

We know that tangential tooth load,

WT= $(\sigma_o \times C_v) \, b.\pi m.y' = (\sigma \times C_v) \, 3pN \times \pi m \times y'...(b=3_pN)$

$= (\sigma_o \times C_v) \, 3 \times p_c \cos\alpha \times \pi m.y' \, (p_N = p_c \cos\alpha)$

$= (\sigma_o \times C_v) \, 3\pi m \cos\alpha \times \pi m.y'.....(p_c = \pi m)$

$$\frac{18600}{m} = 56 \left(\frac{15}{15+1.885 \, m} \right) 3 \, \pi m \times \cos 30° \times \pi m \times 0.129$$

Solving this equation by hit and trial method, we find that,

m = 5.5 say 6 mm

Pitch diameter of the gear: We know that the pitch diameter of the gear,

D_G= m x T_G=6 x 24 = 144mm

Face width: It is given that the face width,

b=3p_N=3$p_c$cos α=3xπm cos α

=3 x π x 6 cos 30°

= 48.98 say 50 mm

End thrust on the gear: We know that end thrust or axial load on the gear,

$$W_A = W_T \tan\alpha = \frac{18600}{m} \times \tan 30° = \frac{18600}{m} \times 0.577 = 1790 \, N$$

4. A 35 kW motor running at 1200 r.p.m. drives a compressor at 780 r.p.m. through a 90° bevel gearing arrangement. The pinion has 30 teeth. The pressure angle of teeth is 14 1/2 °. The wheels are capable of withstanding a dynamic stress, $\sigma_w = 140 \left(\dfrac{280}{280+v} \right)$

MPa, where v is the pitch line speed in m/min. The form factor for teeth may be taken as $0.124 - 0.686/E$, where T_E is the number of teeth equivalent of a spur gear. The face width may be taken as 1/4 of the slant height of pitch cone. Let us determine the module pitch, face width, addendum, dedendum, outside diameter and slant height for the pinion.

Solution:

Given:

$P = 35 \text{ kW} = 35 \times 10^3 \text{ W.}$

$N_p = 1200 \text{ r.p.m.}$

$N_G = 780 \text{ r.p.m.}$

$\theta_s = 90°$

$T_p = 30$

$\varphi = 14\ 1/2°$

$b = L / 4$

Module and face width for the pinion:

Let,

m = Module in mm

b = Face width in mm

= L / 4 ... (Given)

D_p = Pitch circle diameter of the pinion.

We know that velocity ratio,

$$\text{V.R.} = \frac{N_P}{N_G} = \frac{1200}{780} = 1.538$$

∴ Number of teeth on the gear,

$$T_G = \text{V.R.} \times T_p = 1.538 \times 30 = 46$$

Since, the shafts are at right angles, therefore pitch angle for the pinion,

$$\theta_{P1} = \tan^{-1}\left(\frac{1}{\text{V.R.}}\right) = \tan^{-1}\left(\frac{1}{1.538}\right) = \tan^{-1}(0.65) = 33°$$

And pitch angle for the gear,

$$\theta_{P2} = 90° - 33° = 57°$$

We know that formative number of teeth for pinion,

$$T_{EP} = T_P.\sec \theta_{P1} = 30 \times \sec 33° = 35.8$$

And formative number of teeth for the gear,

$$T_{EG} = T_G.\sec \theta_{P2} = 46 \times \sec 57° = 84.4$$

Tooth form factor for the pinion,

$$y'_P = 0.124 - \frac{0.686}{T_{EP}} = 0.124 - \frac{0.686}{35.8} = 0.105$$

And tooth form factor for the gear,

$$y'_G = 0.124 - \frac{0.686}{T_{EG}} = 0.124 - \frac{0.686}{84.4} = 0.116$$

Since the allowable static stress (σ_o) for both the pinion and gear is same (i.e., 140 MPa or N/mm²) and y'_P is less than y'_G, therefore the pinion is weaker. Thus the design should be based upon the pinion.

We know that the torque on the pinion,

$$T = \frac{P \times 60}{2\pi N_P} = \frac{35 \times 10^3 \times 60}{2\pi \times 1200} = 278.5\,N-m = 278\,500\,N-mm$$

∴ Tangential load on the pinion,

$$W_T = \frac{2T}{D_P} = \frac{2T}{m.T_P} = \frac{2 \times 278\,500}{m \times 30} = \frac{18567}{m}\,N$$

We know that pitch line velocity,

$$v = \frac{\pi D_P.N_P}{1000} = \frac{\pi m.T_P.N_P}{1000} = \frac{\pi m \times 30 \times 1200}{1000}\,m/min$$
$$= 113.1\,m/min$$

∴ Allowable working stress,

$$\sigma_w = 140\left(\frac{280}{280+v}\right) = 140\left(\frac{280}{280+113.1\,m}\right)\,MPa\,or\,N/mm^2$$

We know that the length of the pitch cone element or slant height of the pitch cone,

$$L = \frac{D_P}{2\sin\theta_{p1}} = \frac{m \times T_P}{2\sin\theta_{p1}} = \frac{m \times 30}{2\sin 33°} = 27.54\,m\,mm$$

Since the face width (b) is 1/4th of the slant height of the pitch cone, therefore

$$b = \frac{L}{4} = \frac{27.54\,m}{4} = 6.885\,m\,mm$$

We know that tangential load on the pinion,

$$W_T = \left(\sigma_{OP} \times C_v\right) b.\pi m.y'_P \left(\frac{L-b}{L}\right)$$

$$= \sigma_w.b.\pi m.y'_P \left(\frac{L-b}{L}\right) \qquad \dots \left(\because \sigma_w = \sigma_{OP} \times C_v\right)$$

or,

$$\frac{18567}{m} = 140 \left(\frac{280}{280+113.1\,m}\right) 6.885\,m \times \pi m \times 0.105 \left(\frac{27.54\,m - 6.885\,m}{27.54\,m}\right)$$

$$= \frac{66780\,m^2}{280+113.1\,m}$$

or, $280 + 113.1\,m = 66\,780\,m^2 \times \dfrac{m}{18\,567} = 3.6\,m^3$

Solving this expression by hit and trial method, we find that

$$m = 6.6 \text{ say } 8 \text{ mm}$$

And,

$$\text{face width, } b = 6.885\,m = 6.885 \times 8 = 55 \text{ mm}$$

Addendum and dedendum for the pinion,

We know that the addendum,

$$a = 1\,m = 1 \times 8 = 8 \text{ mm}$$

And dedendum,

$$d = 1.2\,m = 1.2 \times 8 = 9.6 \text{ mm}$$

Outside diameter for the pinion:

We know that the outside diameter for the pinion,

$$D_o = D_p + 2\,a \cos \theta_{P_1}$$

$$= m.T_p + 2\,a \cos \theta_{P_1} \dots (\because DP = m.TP)$$

$$= 8 \times 30 + 2 \times 8 \cos 33°$$

$$= 253.4 \text{ mm}$$

Slant height:

We know that the slant height of the pitch cone,

$$L = 27.54 \text{ m} = 27.54 \times 8 = 220.3 \text{ mm}$$

5. A pair of cast iron bevel gears connects two shafts at right angles. The pitch diameters of the pinion and gear are 80 mm and 100 mm respectively. The tooth profiles of the gears are of 14 1/2° composite form. The allowable static stress for both the gears is 55 MPa. If the pinion transmits 2.75kW at 1100 r.p.m., let us determine the module and number of teeth on each gear from the stand-point of strength and check the design from the standpoint of wear. Take the surface endurance limit as 630 MPa and modulus of elasticity for cast iron as 84 kN/mm².

Solution:

Given:

$$\theta_s = 90°$$

$$D_P = 80 \text{ mm} = 0.08 \text{ m}$$

$$D_G = 100 \text{ mm} = 0.1 \text{ m}$$

$$\varphi = 1214°$$

$$\sigma_{OP} = \sigma_{OG} = 55 \text{ MPa} = 55 \text{ N/mm}^2$$

$$P = 2.75 \text{ kW} = 2750 \text{ W}$$

$$N_P = 1100 \text{ r.p.m.}$$

$$\sigma_{es} = 630 \text{ MPa} = 630 \text{N/mm}^2$$

$$E_P = E_G = 84 \text{ kN/mm}^2 = 84 \times 10^3 \text{N/mm}^2$$

Module:

Let m= Module in mm.

$$\theta_{P1} = \tan^{-1}\left(\frac{1}{V.R}\right) = \tan^{-1}\left(\frac{D_P}{D_G}\right) = \tan^{-1}\left(\frac{80}{100}\right) = 38.66°$$

Since, the shafts are at right angles, therefore pitch angle for the pinion,

And pitch angle for the gear,

$$\theta_{P2} = 90° - 38.66° = 51.34°$$

We know that formative number of teeth for pinion,

$$T_{EP} = T_P . \sec \theta_{P1} = \frac{80}{m} \times \sec 38.66° = \frac{102.4}{m} \qquad ...(\because T_P = D_P / m)$$

And formative number of teeth on the gear,

$$T_{EG} = T_G . \sec \theta_{P2} = \frac{100}{m} \times \sec 51.34° = \frac{160}{m} \qquad ...(\because T_G = D_G / m)$$

Since both the gears are made of the same material, therefore pinion is the weaker. Thus, the design should be based upon the pinion. We know that tooth form factor for the pinion having 14 1/2° composite teeth,

$$y'_p = 0.124 - \frac{0.684}{T_{EP}} = 0.124 - \frac{0.684 \times m}{102.4}$$

$$= 0.124 - 0.006\,68\,m$$

And pitch line velocity,

$$v = \frac{\pi D_P . N_P}{60} = \frac{\pi \times 0.08 \times 1100}{60} = 4.6\,m/s$$

Taking velocity factor,

$$C_v = \frac{6}{6+v} = \frac{6}{6+4.6} = 0.566$$

We know that length of the pitch cone element or slant height of the pitch cone,

$$*L = \sqrt{\left(\frac{D_G}{2}\right)^2 + \left(\frac{D_P}{2}\right)^2} = \sqrt{\left(\frac{100}{2}\right)^2 + \left(\frac{80}{2}\right)^2} = 64\ mm$$

* The length of the pitch cone element (L) may also be obtained by using the relation,

$$L = D_P/2 \sin \theta_{P1}$$

Assuming the face width (b) as 1/3rd of the slant height of the pitch cone (L), therefore

$$b = L / 3 = 64 / 3 = 21.3 \text{ say } 22 \text{ mm}$$

We know that torque on the pinion,

$$T = \frac{P \times 60}{2\pi \times N_P} = \frac{2750 \times 60}{2\pi \times 1100} = 23.87\,N - mm$$

∴ Tangential load on the pinion,

$$W_T = \frac{T}{D_P/2} = \frac{23870}{80/2} = 597\,N$$

We also know that tangential load on the pinion,

$$W_T = (\sigma_{OP} \times C_v)\,b \times \pi m \times y'_P \left(\frac{L-b}{L}\right)$$

or, $579 = (55 \times 0.566)\,22 \times \pi m\,(0.124 - 0.00\,668\,m)\left(\dfrac{64-22}{64}\right)$

$$= 1412\,m\,(0.124 - 0.006\,68\,m)$$

$$= 175\,m - 9.43\,m^2$$

Solving this expression by hit and trial method, we find that:

m = 4.5 say 5 mm

Number of teeth on each gear:

We know that number of teeth on the pinion,

$T_P = D_P/m = 80/5 = 16$

And number of teeth on the gear,

$T_G = D_G/m = 100 / 5 = 20$

Checking the gears for wear:

We know that the load-stress factor,

$$K = \frac{(\sigma_{es})^2 \sin \phi}{1.4}\left[\frac{1}{E_P} + \frac{1}{E_G}\right]$$

$$= \frac{(630)^2 \sin 14\,\frac{1}{2}^{\circ}}{1.4}\left[\frac{1}{84\times 10^3} + \frac{1}{84\times 10^3}\right] = 1.687$$

and ratio factor, $Q = \dfrac{2\,T_{EG}}{T_{EG} + T_{EP}} = \dfrac{2\times 160/m}{160/m + 102.4/m} = 1.22$

∴ Maximum or limiting load for wear,

$$W_w = \frac{D_P.b.Q.K}{\cos\theta_{P1}} = \frac{80\times 22\times 1.22\times 1.687}{\cos 38.66^{\circ}} = 4640\,N$$

Since the maximum load for wear is much more than the tangential load (W_T), there-fore the design is satisfactory from the consideration of wear.

6. Let us design a worm gear. A 2kW power is applied to a worm shaft at 720mm. The worm is of quadruple start with 50mm as pitch circle diameter. The worm gear has 40teeth with 5mm module. The pressure angle in the diametric plane is 20°. Also let us determine:

- The lead angle of the worm,

- Velocity ratio,

- Center distance. And also, calculate efficiency of the worm gear drive and power lost in friction.

Solution:

Given:

$\qquad$ P = 2 KW,

$\qquad$ n_1 = 720 rpm,

$\qquad$ d = 50 mm,

$\qquad$ Z = 40; m = 5 mm,

$\qquad$ $\alpha = 20°$.

$$q = \frac{d}{m} = 10$$

Lead angle of worm, $\gamma = \tan^- \left(\dfrac{Z}{q} \right)$

Assume, Number of stacks, Z = 3

$\qquad$ $\therefore \gamma = 16.69°$

$$i = \frac{z}{Z} = \frac{40}{3} 13.33$$

Center distance,

$$a = \left(\frac{z}{q} + 1 \right) \sqrt{\left[\frac{540}{\frac{z}{q} [\sigma_c]} [M_t] \right]}$$

$$[M_t] = \frac{1.3 \times 97420 \times 2}{720} = 351.79 \, \text{kgf cm.}$$

$[\sigma_c]$ = 5000 kgf/cm².

∴ a = 3.176 cm

a = 10 cm

Efficiency of the drive,

Sliding velocity, $v = \dfrac{\pi \times 50 \times 10^{-3} \times 720}{60}$

V = 1.88 *m/s*

μ = 0.04

$\therefore \mu = \dfrac{\tan\gamma}{\tan(\gamma+\rho)} = 84.72\%$

Power lost due to friction,

H_g = (1 – 2) P

= 1.69 KW

6.4   Methods of Avoiding Interference

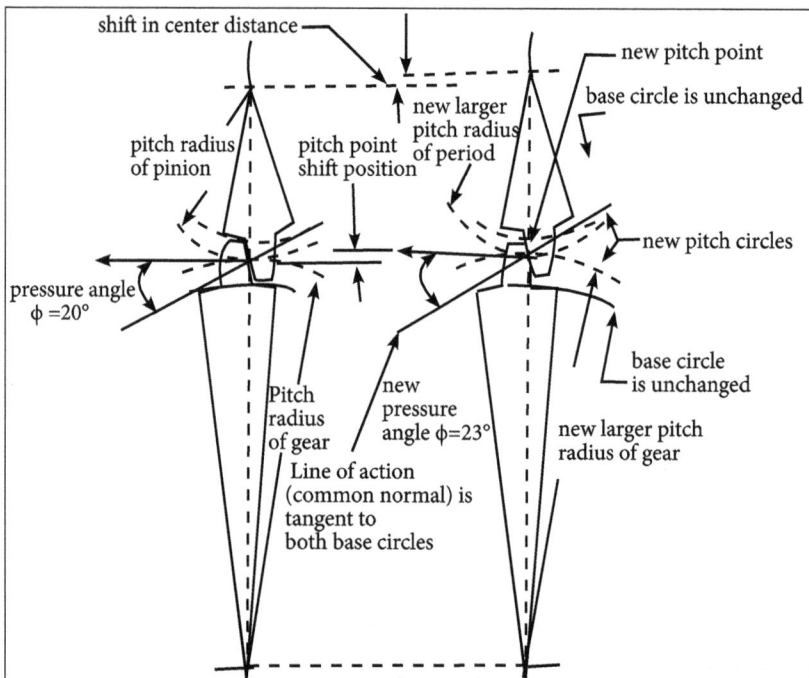

Method of elimination of interference in spur gears.

In certain spur designs if interference exists, it can be overcome by:

- Removing the cross hatched tooth tips i.e., using stub teeth.

- Increasing the number of teeth on the mating pinion.

- Increasing the pressure angle.

- Tooth profile modification or profile shifting.

- Increasing the Centre distance as illustrated in figure. For a given gear, the interference can also be eliminated by increasing the centre distance.

Minimum No. of Teeth on Pinion to Avoid Interference for a Given Gear

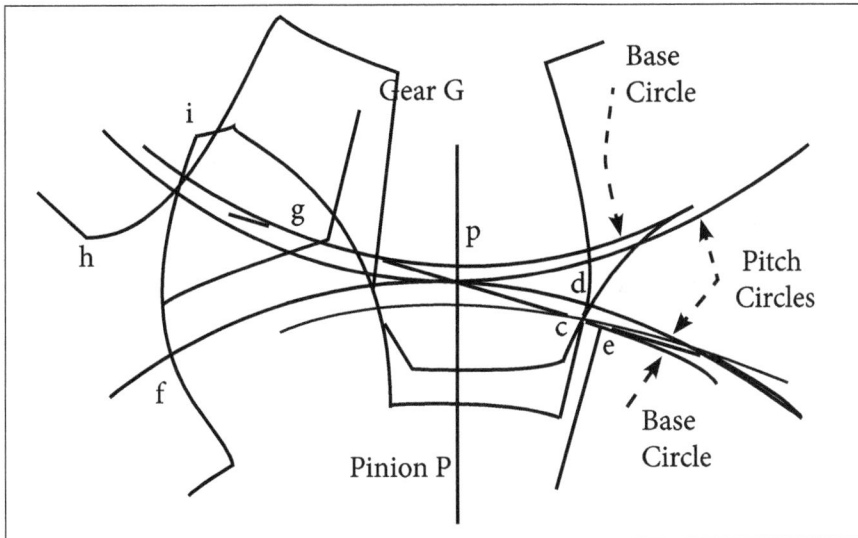

(a) Gears in mesh showing portion of the tooth of pinion digging into the gear tooth on the left.

- Above the figure (a), the involute profile does not exist beyond base circle. When the pinion rotates clockwise, first and last point of contacts are e and g where the line of action is tangent to the base circles.

- Any part of the pinion tooth face extending beyond a circle through g interferes with gear flank as shown at i.

- The interference limits the permissible length of addendum. As the diameter of the pinion is reduced, the permissible addendum of larger gear becomes smaller.

- Let the addendum height be k times the module i.e., km. From the figure (b),the maximum gear addendum circle radius is given equation,

$$AE = r_2 + km = \sqrt{AG^2 + GE^2} \quad ...(i)$$

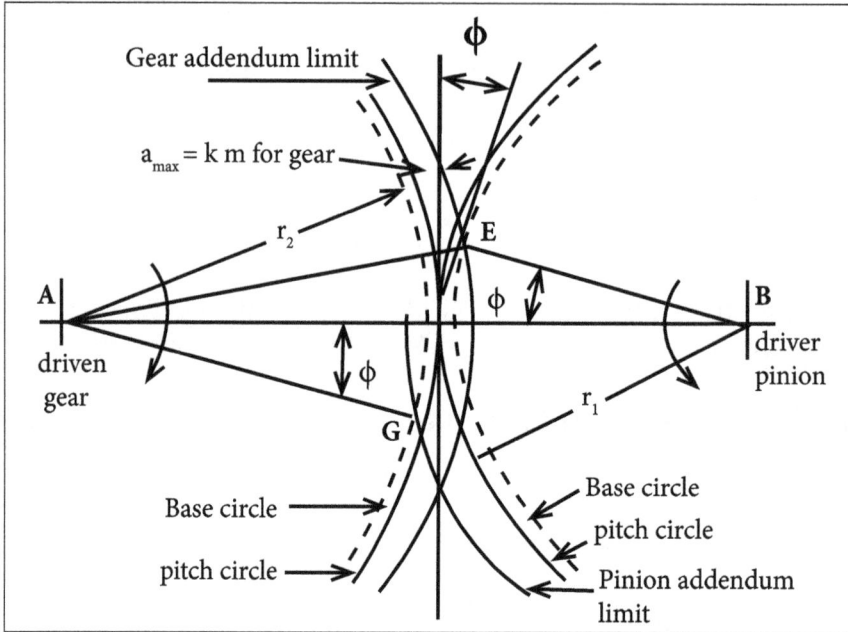

(b) Pinion and gear meshing shown by equivalent circles.

But $AG = r_2 \cos\varphi$ and $GE = (r_1 + r_2)\sin\varphi$

Substituting in equation (i) and simplifying

$$r_2 + km = \sqrt{r_2^2 \cos^2\varphi + (r_1 + r_2)^2 \sin^2\varphi} \quad \ldots\text{(ii)}$$

Substituting $r_1 = mz_1$ and $r_2 = mz_2$ in equation (ii) and rearranging.

$$z_1^2 + 2z_1z_2 = \frac{4k(z_2 + k)}{\sin^2\varphi} \quad \ldots\text{(iii)}$$

For a rack and pinion, $z_2 = \infty$ and the equation (iii) reduces to,

$$Z_1 = \frac{2k}{\sin^2\varphi}$$

For full depth gears (i.e., $k = 1$) engaging with rack, minimum teeth on the pinion to avoid interference is given by,

$z_1 = 31.9 = 32$ for $14.5°$ pressure angle.

$z_1 = 17.097 = 17$ for $20°$ pressure angle.

$z_1 = 13.657 = 14$ for $22.5°$ pressure angle rounded to integer value.

The equation (iii) indicates that the minimum number of teeth on pinion permissible and it depends on the gear ratio and pressure angle.

From the practical consideration it is observed that rack gear generation and hobbing process for lower value than the one given earlier, a little undercutting takes place and the strength of the gear is not affected. Hence, corresponding minimum number of teeth are 27, 14 and 12 for 14.50, 200 and 22.50 instead of 32, 17 and 14.

6.4.1 Backlash

Backlash is the maximum distance or angle through which any part of a mechanical system may be moved in one direction without applying appreciable force or motion to the next part in a mechanical sequence.

An example, in the context of gears and gear trains, is the amount of clearance between mated gear teeth. It can be seen when the direction of movement is reversed and the slack or lost motion is taken up before the reversal of motion is complete. Another example is in a valve train with mechanical tappets, where a certain range of lash is necessary for the valves to work properly.

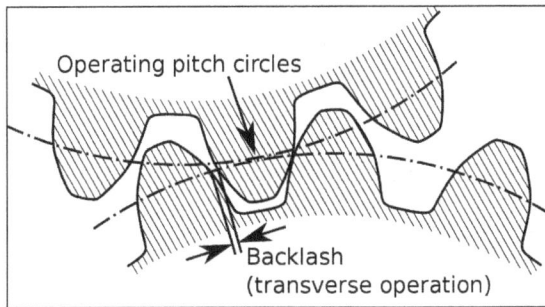

Backlash.

Depending on the application, backlash may or may not be desirable. It is unavoidable for nearly all reversing mechanical couplings, although its effects can be negated or compensated for. In many applications, the theoretical ideal would be zero backlash, but in actual practice some backlash must be allowed to prevent jamming. Reasons for the presence of backlash include allowing for lubrication, manufacturing errors, deflection under load and thermal expansion.

Types of backlashes:

Circumferential backlash (j_t): Circumferential Backlash is the length of arc on the pitch circle. The length is the distance the gear is rotated until the meshed tooth flank makes contacts while the other mating gear is held stationary.

Normal backlash (j_n): The minimum distance between each meshed tooth flank in a pair of gears, when it is set so the tooth surfaces are in contact.

Angular backlash (j_θ): The maximum angle that allows the gear to move when the other mating gear is held stationary.

Radial backlash (j_r): The radial Backlash is the shrinkage (displacement) in the stated center distance when it is set so the meshed tooth flanks of the paired gears get contact each other.

Axial backlash (j_x): The axial backlash is the shrinkage (displacement) in the stated center distance when a pair of bevel gears is set so the meshed tooth flanks of the paired gears contact each other.

6.5   Comparison of Involute and Cycloidal Teeth

In-volute Tooth Profile	Cycloidal Profile
Variation in center distance does not affect the velocity ratio.	The center distance should not vary.
Pressure angle remains constant throughout the teeth.	Pressure angle varies. It is zero at the pitch point and maximum at the start and end of engagement.
Interference occurs.	No interference occurs.
Weaker teeth.	Stronger teeth.

Cycloidal Teeth

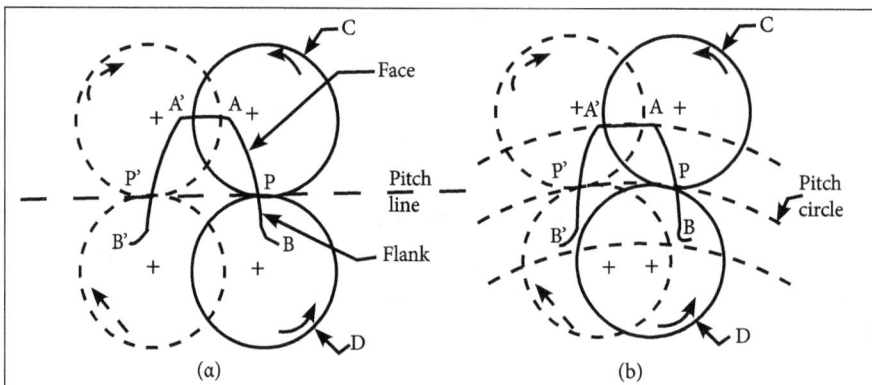

Cycloidal teeth.

The curve traced by a point on the circumference of a circle rolling on a fixed straight line without slipping is called a cycloid. The curve traced by a point on the circumference of a circle when it rolls without slipping on the outside of a fixed circle, is known as epicycloid. If a circle rolls without slipping on the inside of a fixed circle, then the curve traced by a point on the circumference of a circle is called hypocycloid.

Involute Teeth

The plane curve generated by a point on a tangent, which rolls on the circle without slipping or by a point on a string wrapped from a reel, is known as Involute of a circle. In relation with toothed wheels it is known as base circle.

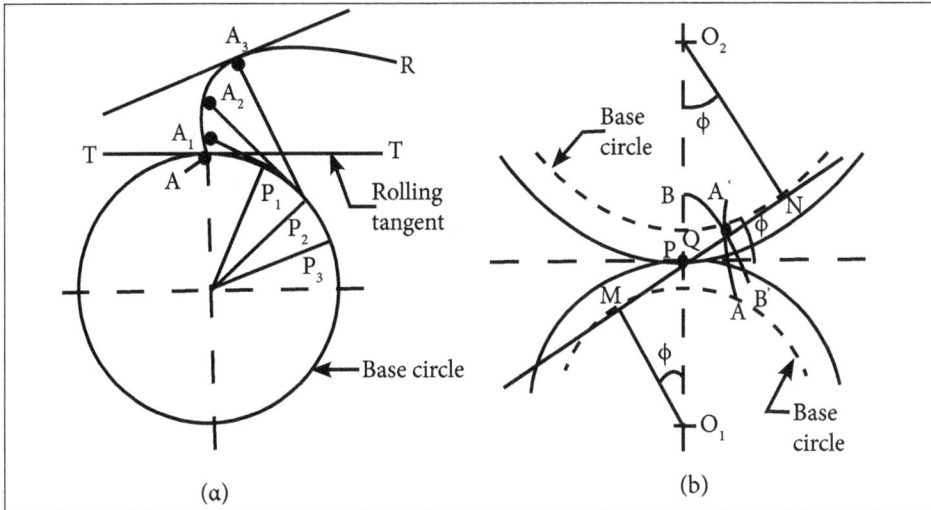

Involute teeth.

Advantages of cycloidal gears: 1) Since, the cycloidal teeth have wider flanks; therefore the cycloidal gears are stronger than the involute gears, for the same pitch. Due to this reason, the cycloidal teeth are preferred especially for cast teeth. 2) In cycloidal gears, the contact takes place between a convex flank and concave surface, where as in involute gears, the convex surface are in contact. This condition results in less wear in cycloidal gears as compared to involute gears. However the difference in wear is negligible. 3) In the cycloidal gears, the interference does not occur at all. Though there are advantages of cycloidal gears but they are out weighted by the greater simplicity and flexibility of the involute gears.

Advantages of involute gears: 1) The most important advantage of the involute gears is that the center distance for a pair of involute gears can be varied within limits without changing the velocity ratio. This is not true for cycloidal gears which require exact center distance to be maintained. 2) In involute gears, the pressure angle, from the start of the engagement of teeth to the end of the engagement, remains constant. It is necessary for smooth running and less wear of gears. But in cycloidal gears, the pressure angle is maximum at the beginning of engagement, reduces to zero at pitch point, starts increasing and again become maximum at the end of engagement. This results in less smooth running of gears. 3) The face and flank of involute teeth are generated by a single curve where as in cycloidal gears, double curves are required for the face and flank respectively. Thus the involute teeth are easy to manufacture than cycloidal teeth. In involute system, the basic rack has straight teeth and the same can be cut with simple tools.

6.5.1 Profile Modification

The figure shows the geometry of spur gear tooth with circular root fillet in which point 'O' is the center of the gear, 'Oy' is the axis of symmetry of the tooth and 'B' is the point where the involute profile starts (from the form circle r_s).

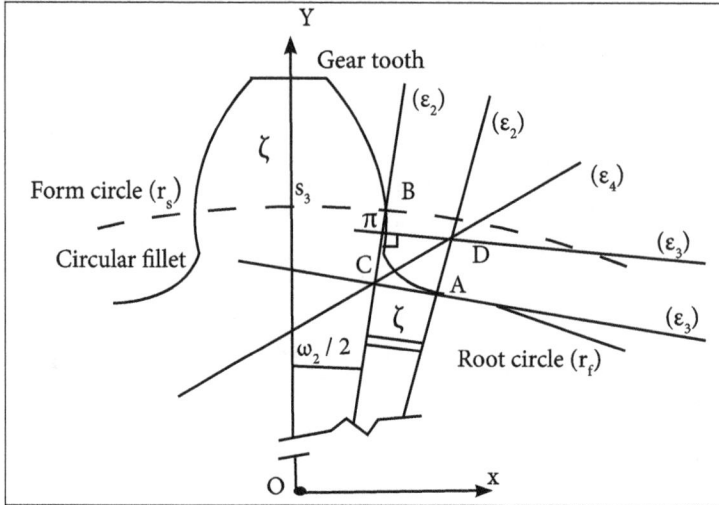

(a) Spur gear geometry of the circular root fillet.

Point 'A' is the point of tangency of the circular fillet with the root circle r_f. Point 'D' lying on (ε_2) identical to 'OA' represents the center of the circular fillet. Line (ε_3) is tangent to the root circle at 'A' and intersects with line (ε_1) at 'C'. The fillet is tangent to the line (ε_1) at point 'E'.

Since it is always $r_s > r_f$, the proposed circular fillet can be implemented without exceptions on all spur gears irrelevant of the number of gear teeth or other manufacturing parameters.

A comparison of the geometrical shape of a tooth of circular root fillet with that of the standard trochoidal root fillet is presented in the figure (b). The geometry of the circular root fillet which coordinates (points A, B, C, D and E) in the figure (a), are obtained using the formulas given below:

$$X_B = r_f \operatorname{Sin}\Omega_s$$

$$Y_B = r_f \operatorname{Cos}\Omega_s$$

$$Y_C = \frac{X_C}{\operatorname{Tan}\Omega_s}$$

$$X_A = r_f \operatorname{Sin}(\zeta + \Omega_s)$$

$$Y_A = r_f \operatorname{Cos}(\zeta + \Omega_s)$$

$$X_E = (OC + CE)\sin\Omega_s$$

$$Y_E = (OC + CE)\cos\Omega_s$$

$$X_D = (r_f + AD)Sin(\zeta + \Omega_s)$$

$$Y_D = (r_f + AD)Cos(\zeta + \Omega_s)$$

$$X_C = rf \frac{Tan\Omega_s}{Sin(\zeta + \Omega_s)Tan\Omega_s + Cos(\zeta + \Omega_s)}$$

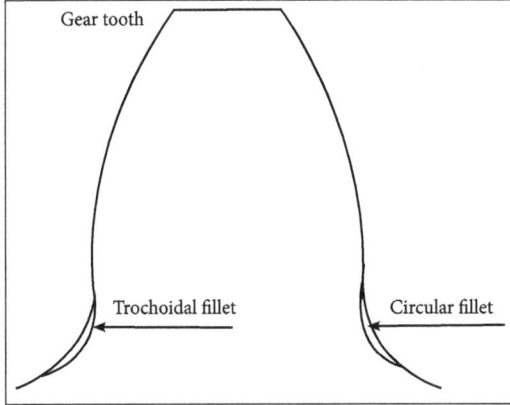

(b) Superposition of circular fillet on a standard tooth.

The remaining portion of the tooth profile between points 'B' and 'E' is a straight line. Angle $\omega_s/2$ that corresponds to the arc $S_s/2$ in the figure (a), is given by the equation (i),

$$\omega_s/2 = \frac{S_s/2}{r_s} = \Omega_s \qquad ..(i)$$

Angle ζ (Figure (a)) takes values between o and ζ_{max} (Equation (ii))

$$\zeta_{max} = \frac{\pi}{Z} - \Omega_s \qquad ...(ii)$$

Gear Trains

7.1 Introduction to Gear Trains

Sometimes, two or more gears are made to mesh with each other to transmit power from one shaft to another. Such a combination is called gear train or train of toothed wheels.

The nature of the train used depends upon the velocity ratio required and the relative position of the axes of shafts. A gear train may consist of spur, bevel or spiral gears.

Types of Gear Trains

Following are the different types of gear trains, de-pending upon the arrangement of wheels:

- Simple gear train
- Compound gear train
- Re-verted gear train
- Epicyclic gear train

In the first three types of gear trains, the axes of the shafts over which the gears are mounted are fixed relative to each other. But in case of epicyclic gear trains, the axes of the shafts on which the gears are mounted may move relative to a fixed axis.

A gear train is two or more gear working together by meshing their teeth and turning both other in a system to generate power and speed. It reduces speed and increases torque. To create large gear ratio, gears are connected together to form gear trains. They often consist of multiple gears in the train.

The most common of the gear train is the gear pair connecting parallel shafts. The teeth of this type can be spur, helical or herringbone. The angular velocity is simply the reverse of the tooth ratio. Any combination of gear wheels employed to transmit motion from one shaft to the other is called a gear train.

The meshing of two gears may be idealized as two smooth discs with their edges touching and no slip between them. This ideal diameter is called the Pitch Circle Diameter (PCD) of the gear.

7.1.1. Simple Gear Trains

It the axes of all gears (or the axes of the shafts on which gears are mounted) remain fixed relative to each other, the gear train is known as simple gear train or ordinary gear train.

In case of a simple gear train, each gear is on a separate shaft as shown in the figure. In figure (a), there are only two gears. Each gear is mounted on the separate shaft. The combination of these two gears is known as simple gear train. If power is transmitted from gear 1 to gear 2, then gear I is driver whereas gear 2 is driven or follower. These two gears rotate in opposite direction.

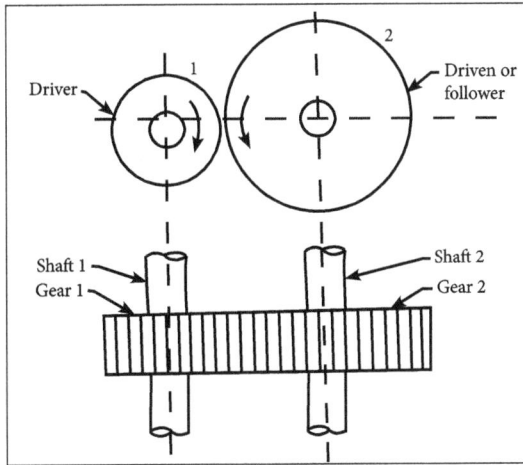

(a) Simple gear train.

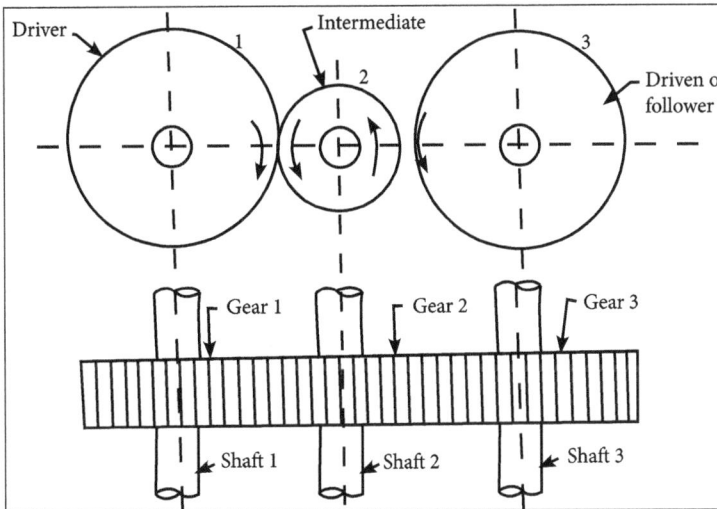

(b) Simple gear train.

In figure (b), there are three gears. The shaft 1 carries only one gear, shaft 2 also carries only one gear and shaft 3 also has only one gear. The combination of these gears, is known as simple gear train, when they are arranged in such a way that power is transmitted from a driving shaft to a driven shaft.

The shaft 1 is a driver whereas the shaft 3 is known as driven shaft or follower. The shaft 2 is called intermediate shaft. It may be noted that when the number of intermediate shafts are odd, the motion of both the shafts or gears (i.e. driver and follower) is like. But if the number of intermediate shafts are even, the motion of follower will be in the opposite direction of the driver.

Speed Ratio and Train Value

When there is only one gear on each shaft, as shown in figure, it is known as simple gear train. The gears are represented by their pitch circles. When the distance between the two shafts is small, the two gears 1 and 2 are made to mesh with each other to transmit motion from one shaft to the other, as shown in figure (a).

Since, the gear 1 drives the gear 2, therefore gear 1 is called the driver and the gear 2 is called the driven or follower. It may be noted that the motion of the driven gear is opposite to the motion of driving gear.

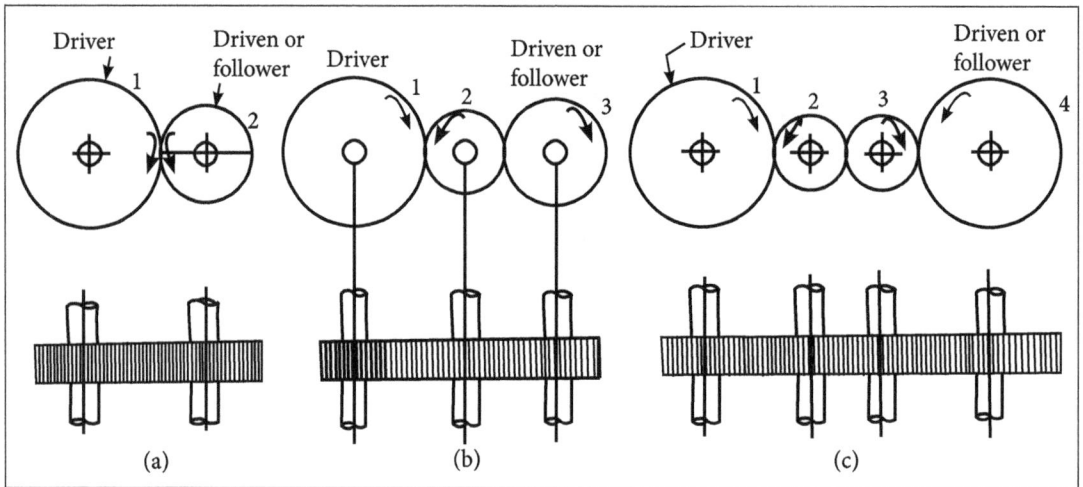

Transmit motion from one shaft to the other.

Let,

N_1 = Speed of gear 1(or driver) in rpm,

N_2 = Speed of gear 2 (or driven or follower) in rpm,

T_1 = Number of teeth on gear 1 and

T_2 = Number of teeth on gear 2.

Since, the speed ratio (or velocity ratio) of gear train is the ratio of the speed of the driver to the speed of the driven or follower and ratio of speeds of any pair of gears in mesh is the inverse of their number of teeth, therefore,

$$\text{Speed ratio} = \frac{N_1}{N_2} = \frac{T_2}{T_1}$$

It may be noted that ratio of the speed of the driven or follower to the speed of the driver is known as train value of the gear train, mathematically,

$$\text{Train value} = \frac{N_2}{N_1} = \frac{T_1}{T_2}$$

The train value is the reciprocal of speed ratio. Sometimes, the distance between the two gears is large.

The motion from one gear to another, in such a case, may be transmitted by either of the following two methods:

- By providing the large sized gear.

- By providing one or more intermediate gears.

A little consideration will show that the former method (i.e. providing large sized gears) is very inconvenient and uneconomical method, whereas the latter method (i.e. providing one or more intermediate gear) is very convenient and economical.

It may be noted that when the number of intermediate gears are odd, the motion of both the gears (i.e. driver and driven or follower).Will be in the direction of drive.

But if the number of intermediate gears is even, the motion of the driven or follower will be in the opposite direction of the drive.

Application:

- To connect gears where a large center distance is required.

- To obtain high speed ratio.

- To obtain desired direction of motion of the driven gear.

7.1.2 Compound Gear Train

A compound gear train may be defined as two or more gear pairs, arranged in series. At least one shaft in a compound gear train should have two gears mounted on it. Figure shows a compound train consisting of only two gear pairs. The gears having number of teeth t_1 and T_1 form the first simple gear pair and the gears having number of teeth t_2 and T_2 form the second gear pair.

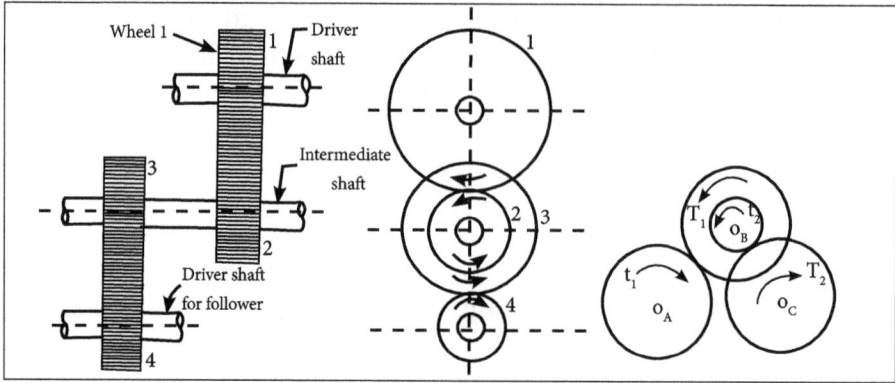

Compound gear train.

The speed of the wheels mounted on the shaft B is the same, i.e. The follower of the first gear pair and the driver of the second gear pair have the same speed. Denoting n and N as the speeds of the driver and follower wheels with subscripts 1or 2, referring to gear pair 1 and 2 respectively,

$$N_1 = n_1 \frac{t_1}{T_1}$$

$$n_2 = N_1$$

$$N_2 = n_2 \frac{t_2}{T_2}$$

$$N_2 = n_1 \frac{t_1}{T_1} \frac{t_2}{T_2}$$

Therefore, the overall gear ratio $R_{1,2}$ for the two simple gear pairs in series,

$$R_{1,2} = \frac{n_1}{N_2} = \frac{T_1 T_2}{t_1 t_2}$$

For a compound gear train consisting of n simple gear pairs, the overall gear ratio $R_{1,2....n}$ will be,

$$R_{1,2,...n} = \frac{T_1 T_2 T_3 ... T_n}{t_1 t_2 t_3 ... t_n}$$

Since $T_1, T_2, ...$ and $t_1, t_2, ...$ are whole numbers, the overall gear ratio of a compound train is given by,

$$R = \frac{U}{u}$$

Where, $U = T_1 T_2 T_3 ... T_n$ and $u = t_1 t_2 t_3 ... t_n$

The whole numbers U and u can be factorized and suitable combinations chosen, for the design of a compound gear train.

Gear Ratio

The gear ratio of a gear train, also known as its speed ratio, is the ratio of the angular velocity of the input gear to the angular velocity of the output gear. The gear ratio can be calculated directly from the numbers of teeth on the gears in the gear train.

$$R = \frac{\omega_A}{\omega_B} = \frac{N_B}{N_A}$$

Where,

ω_A, ω_B = Angular velocity of input and output gear respectively,

N_a, N_b = Number of teeth on the input gear and output gear respectively.

Parallel Axis Gear Trains

When all the axis of the shaft of the gears are in the same plane , it is called as parallel axis gear trains. Very often a single gear pair transferring motion between two parallel shafts cannot meet the needed transmission ratio requirement.

For example, if this requirement is 10:1, then the diameter of the gear must be 10 times larger the diameter of the pinion. This is usually unacceptable from the design specifications concerning product size.

- The solution is achieved by arranging a series of gear pairs. Such series is known as gear train.

- A train may comprise different type of gears: spur, bevel, helical and worm. The gears in a train are functionally in series with each other.

- If a system comprises a few trains, they are functionally in parallel with each other. Usually a system of gears arranged physically in one case (box), whether in series or in parallel, is known as transmission box.

7.2　Epicyclic Gear Trains

The gear trains arranged in such a manner that one or more of their members moves upon and around another member are known as epicyclic gear train. The epicyclic gear train may be simpler or compound train.

Epicyclic Gear Train

Epicyclic means one gear revolving upon and around another. The design involves planet and sun gears as one orbit the other like a planet around the sun. Here is a picture of a typical gear box. This design can produce large gear ratios in a small space and are used on a wide range of applications from marine gear boxes to electric screwdrivers.

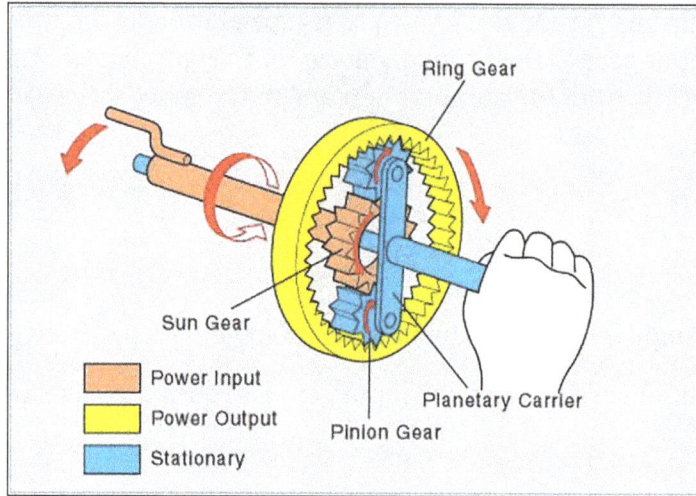

Epicyclic gear train.

Uses

- Back Gear of Lathe.
- Differential Gears of Automobile.
- Hoists, Wrist Watches, etc.

Differential

An automobile while running the inner wheel turns over a smaller radius while outer moves along a larger radius thus larger wheel travels a longer distance then inner wheel hence, there should be difference in speed of the two wheels to accommodate easy turning. Differential is an arrangement of gears which causes this difference in speed between the two vehicles.

The propeller shaft is driven by the engine through the gear box, on which a bevel gear P (known as pinion) is keyed through universal coupling. Bevel gear P meshes with gear Q (known as crown gear), which is free to rotate about the axle A.

Two equal gears R and S are mounted on two separate rod A and A' of the rear axles respectively. These gears, in turn mesh with equal pinions T and U which can rotate freely on the spindle provided on the arm attached to gear Q.

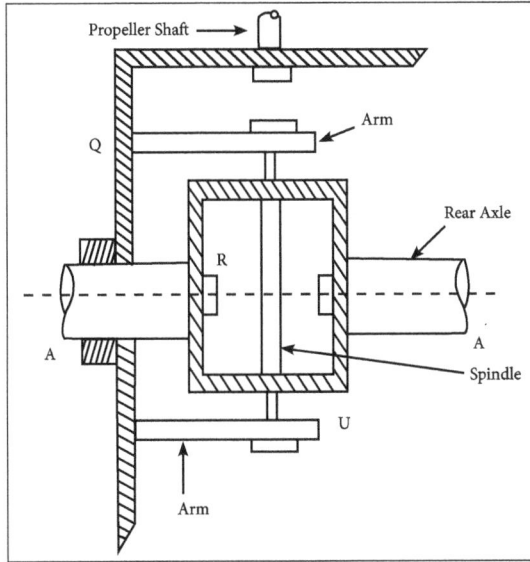

Automobile

When the automobile runs on a straight path, the gears Q and S must rotate together. These gears are rotated through the spindle on the gear Q. The gears T and U do not rotate on the spindle, while taking a turn, outer wheel has greater speed than inner wheel, the gears T and U start rotating about the spindle axis and at the same time revolve about the axle axis.

Due to this epicyclic effect, the speed of inner rear wheel decreases by an amount equal to the increase in speed of outer real wheel.

7.2.1 Algebraic and Tabular Methods of Finding Velocity Ratio of Epicyclic Gear Trains

The following two methods may be used for finding out the velocity ratio of an epicyclic gear train:

- Tabular method
- Algebraic method

These methods are discussed in detail, as follows:

Tabular Method

Consider an epicyclic gear train as shown in figure below.

Let,

T_A = Number of teeth on gear A.

T_B = Number of teeth on gear B.

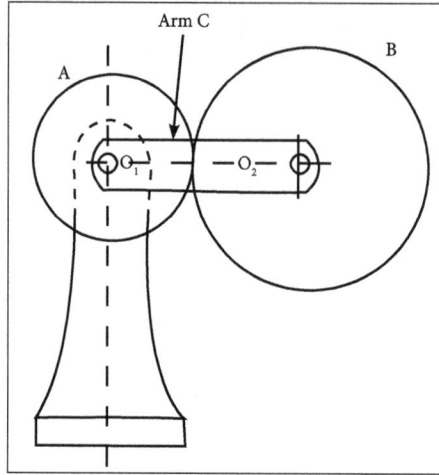

Epicycle gear train.

Let us suppose that the arm is fixed. Therefore the axes of both the gears are also fixed relative to H. When the gear A makes one revolution anticlockwise, the gear B will make T_A/T_B revolutions clockwise. Assuming the anticlockwise rotation as positive and clockwise as negative, we know that when gear A makes + 1 revolution, then the gear B will make $(-T_A/T_B)$ revolutions.

Secondly, if the gear A makes + x revolutions, then the gear B will make $-x \times T_A/T_B$ revolutions. This statement is entered in the second row of the below table. In other words, multiply the each motion (entered in the first row) by x.

Thirdly, each element of an epicyclic train is given +y revolutions and entered in the third row. Finally, the motion of each element of the gear train is added up and entered in the fourth row. We Know that, $N_B/N_A = T_A/T_B$. Since, $N_A = 1$ revolution, therefore, $N_B = T_A/T_{B'}$

Step No.	Condition of motion	Revolution of elements		
		Arm C	Gear A	Gear B
1.	Arm fixed-gear A rotates through +1 revolution i.e., 1 revolution anticlockwise	0	+1	$-\dfrac{T_A}{T_B}$
2.	Arm fixed-gear A rotates through + x revolutions	0	+x	$-x \times \dfrac{T_A}{T_B}$
3.	Add +y revolution to all elements	+y	+y	+y
4.	Total motion	+y	x+y	$y - x \times \dfrac{T_A}{T_B}$

A little consideration will show that when two conditions about the motion of rotation of any two elements are known, then the unknown speed of the third element may be obtained by substituting the given data in the third column of the fourth row.

Algebraic Method

In this method, the motion of each element of the epicyclic train relative to the arm is set down in the form of equations. The number of equations depends upon the number of elements in the gear train. But the two conditions are usually, supplied in any epicyclic train viz. some element is fixed and the other has specified motion. These two conditions are sufficient to solve all the equations and hence, to determine the motion of any element in the epicyclic gear train.

Let the arm C be fixed in an epicyclic gear train as shown in figure below. Therefore, speed of the gear A relative to the arm C,

$= N_A - N_C$

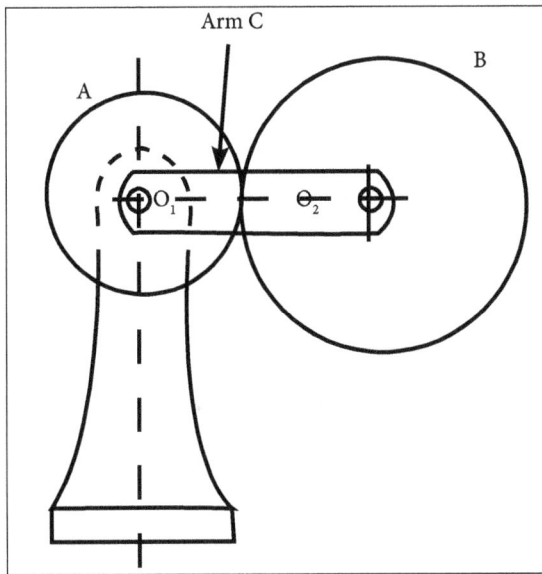

Epicycle Gear Train.

Speed of the gear B relative to the arm C,

$= N_B - N_C$

Since, the gear A and B are meshing directly, therefore they will revolve in opposite directions,

$$\frac{N_B - N_C}{N_A - N_C} = -\frac{T_A}{T_B}$$

Since, the arm C is fixed, therefore its speed, $N_C = 0$,

$$\frac{N_B}{N_A} = -\frac{T_A}{T_B}$$

If the gear A is fixed, then $N_A = 0$,

$$\frac{N_B - N_C}{0 - N_C} = -\frac{T_A}{T_B} \qquad \text{or} \qquad \frac{N_B}{N_c} = 1 + \frac{T_A}{T_B}$$

Problems

1. In an epicyclic gear train shown in figure, the arm A is fixed to the shaft S. The wheel B having 100 teeth rotates freely on the shaft S. The wheel F having 150 teeth is driven separately. If the arm rotates at 200 rpm and wheel F at 100 rpm in the same direction. Let us find (a) number of teeth on the gear C and (b) speed of wheel B.

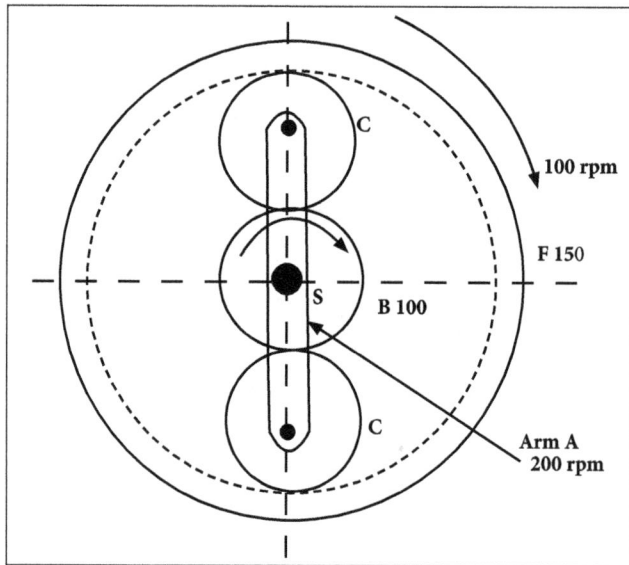

Solution:

Given data:

$T_B = 100;$

$T_F = 150;$

$N_A = 200 \text{rpm};$

$N_F = 100 \text{rpm}.$

Formula to be used:

$$\therefore \qquad -\frac{T_B}{T_F} = \frac{N_F - N_A}{N_B - N_A}$$

$$r_F = r_B + 2r_c$$

The number of teeth on the gears is proportional to the pitch circles,

$r_F = r_B + 2r_c$

$T_F = T_B + 2T_C$

$150 = 100 + 2 \times T_c$

$T_c = 25 \rightarrow$ Number of teeth on gear C.

The gear B and gear F rotates in the opposite directions.

$\therefore$ Train value $= -\dfrac{T_B}{T_F}$

also, $\quad TV = \dfrac{N_L - N_{Arm}}{N_F - N_{Arm}} = \dfrac{N_F - N_A}{N_B - N_A}$

$\therefore \qquad -\dfrac{T_B}{T_F} = \dfrac{N_F - N_A}{N_B - N_A}$

$-\dfrac{100}{150} = \dfrac{100 - 200}{N_B - 200} \qquad \Rightarrow \qquad N_E = 350$

The Gear B rotates at 350 rpm in the same direction of gears F and Arm A.

2. An epicyclic gear train is shown in figure. The Input is given to the gear A which has 24 teeth. Gear wheels B and C constitute a compound planet having 30 and 18 teeth respectively. If all the gears are of the same pitch, let us find the speed ratio of the gear train assuming the gear wheel E is fixed.

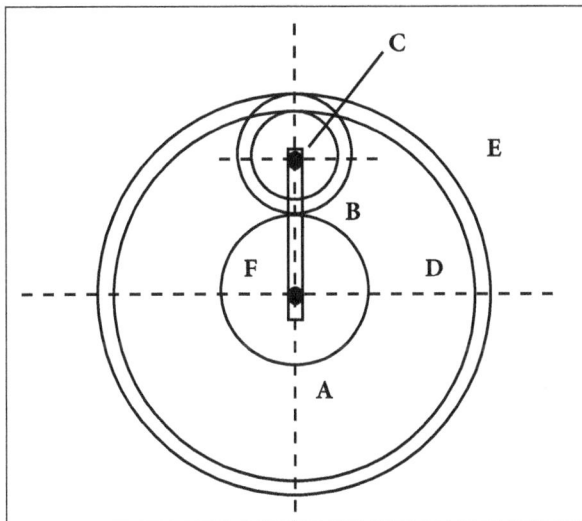

Solution:

Given data:

$$T_a = 24 \text{ teeth}; \quad T_B = 30 \text{ teeth}; \quad T_C = 18 \text{ teeth}.$$

Step No.	Conditions of Motion	Revolutions of Elements				
		Arm F	Gear A	Compound Gear B-C	Gear D	Gear E
1.	Arm F fixed; Gear A rotates through +1 review. i.e., (+1 review anticlockwise).	0	+1	$-\dfrac{T_A}{T_B}$	$-\dfrac{T_A}{T_B} \times \dfrac{T_C}{T_D}$	$-\dfrac{T_A}{T_B} \times \dfrac{T_B}{T_E} = -\dfrac{T_A}{T_E}$
2.	Arm F fixed; sun gear A rotates through +x revolutions.	0	+x	$-x\dfrac{T_A}{T_B}$	$-x\dfrac{T_A}{T_B} \times \dfrac{T_C}{T_D}$	$-x\dfrac{T_A}{T_E}$
3.	Add +Y revolutions to all elements.	+Y	+Y	+Y	+Y	+Y
4.	Total Motion	+Y	x+Y	$Y - x\dfrac{T_A}{T_B}$	$Y - x\dfrac{T_A}{T_B} \times \dfrac{T_C}{T_D}$	$Y - x\dfrac{T_A}{T_E}$

Ratio of the Reduction Gear:

$$\text{Ratio of the reduction} = \frac{\text{Speed of the Input Gear}}{\text{Speed of the Output Gear}} = \frac{N_A}{N_D}$$

From the geometry of the figure, we can write,

$$\frac{d_E}{2} = \frac{d_A}{2} + d_B$$

$$\frac{d_D}{2} = \frac{d_A}{2} + \frac{d_B}{2} + \frac{d_C}{2} \quad \text{(or)} \quad d_D = d_A + d_B + d_C$$

Where, d_A, d_B, d_c, d_D and d_E are the pitch circle.

The number of teeth is proportional to the pitch circle diameter. So,

$$\frac{T_E}{2} = \frac{T_A}{2} + T_B \quad \Rightarrow \quad \frac{T_E}{2} = \frac{24}{2} + 30$$

$$T_e = 84 \text{ and } T_D = T_A + T_B + T_c$$

$$T_d = 24 + 30 + 18 = 72$$

The given conditions are:

(i) Gear wheel E is fixed. So,

$$Y - x\frac{T_A}{T_E} = 0$$

$$Y - x \times \frac{24}{84} = 0 \quad \text{or} \quad Y = \frac{2x}{7}$$

Speed of input gear A, $N_A = x + Y$ and Speed of output gear D,

$$N_D = Y - x \times \frac{T_A}{T_B} \times \frac{T_C}{T_D}$$

$$N_D = Y - x \times \frac{24}{30} \times \frac{18}{72}$$

$$N_d = Y - 0.2\, x.$$

Ratio of reduction gear,

$$\frac{N_S}{N_D} = \frac{x + Y}{Y - 0.2x} = \frac{x + \dfrac{2x}{7}}{\dfrac{2x}{7} - 0.2x}$$

$$\frac{N_S}{N_D} = \frac{x\left(1 + \dfrac{2}{7}\right)}{x\left(2/7 - 0.2\right)} = 15$$

Result:

The speed ratio,

$$\frac{N_S}{N_D} = 15$$

3. The pitch circle diameter of the annular gear in the epicyclic gear train in figure is

425 mm and the module is 5 mm. When the annular gear 3 is stationary, the spindle A makes one revolution in the same sense as the sun gear 1 for every 6 revolutions of the driving spindle carrying the sun gear.

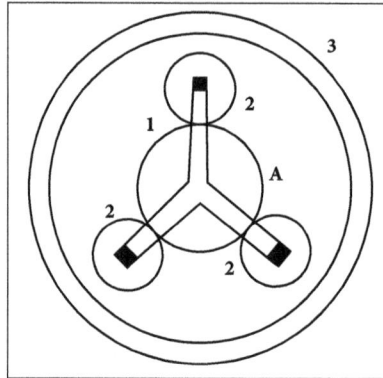

All the planet gears are of the same size. Let us determine the number of teeth on all gears.

Solution:

Step No.	Conditions of Motion	Spider A	Revolutions of Elements		
			Sun Wheel 1	Planet Wheel 2	Annular Gear 3
1.	Spider A fixed; Sun Wheel 1 rotates through +1 revolve	0	+1	$-\dfrac{T_1}{T_2}$	$-\dfrac{T_1}{T_2} \times \dfrac{T_2}{T_3} = \dfrac{-T_1}{T_3}$
2.	Spider A fixed; Sun Wheel 1 rotates through +x revolution	0	+x	$-x\dfrac{T_1}{T_2}$	$-x\dfrac{T_1}{T_3}$
3.	Add +y revolution to all element	+y	+y	+y	+y
4.	Total Motion	+y	x + y	$y - x\dfrac{T_1}{T_2}$	$y - x\dfrac{T_1}{T_3}$

$$y = +1; \quad x + y = +6$$

$$x = 6 - 6 = 6 - 1 = 5$$

Given gear A is Stationary,

$$y - x\frac{T_1}{T_3} = 0$$

$$1 - 5\frac{T_1}{T_3} = 0$$

$$T_3 = \frac{425}{5}$$

$$T_3 = 85$$

$$1 - 5\left(\frac{T_1}{85}\right) = 0$$

$$1 - \frac{5T_1}{85}$$

$$T_1 = 17$$

Where,

$$d_3 = 2d_2 + d_1$$

$$T_3 = 2T_2 + T_1$$

$$85 = 2T_2 + 17$$

$$T_2 = 51$$

Tooth Load and Torque Calculations in Epicyclic Gear Trains

Assuming the different gears in the epicyclic gear train to move with uniform speed (i.e. accelerations are zero), then sum total of the torques must be zero, i.e.

Let,

t_i = Input torque or driving torque having wheel speed, coi.

t_o = Output torque or resisting torque having wheel speed, coo.

t_h = Holding torque or braking torque which makes the corresponding wheel speed, coh zero.

Or $t_i + t_o + t_h = 0$.

If the friction losses of the teeth during mesh and the frictional losses at bearings neglected, then the total energy must be equal to zero,

$$\sum t.\omega = 0$$

or, $t_i \omega_i + t_o \omega_o + t_h \omega_h = 0$

or, $t_i \omega_i + t_o \omega_o = 0$ as $\omega_h = 0$

or, $t_o = -t_i \dfrac{\omega_i}{\omega_o}$

and $t_h = -(t_i + t_o) = -\left(t_i - t_i \dfrac{\omega_i}{\omega_o}\right) = t_i\left(\dfrac{\omega_i}{\omega_o} - 1\right)$

$\dfrac{\omega_i}{\omega_o} = 1 + \dfrac{T_A}{T_S}$ for first gear ratio, thus,

$t_o = t_C = -t_S\left(1 + \dfrac{T_A}{T_S}\right)$ (-v sign is for the movement in the opposite direction)

And $t_h = -t_s \dfrac{T_A}{T_S}$

Problems

1. An epicyclic gear consists of three gears A, B and C as shown in Figure. The gear A has 72 internal teeth and gear C has 32 external teeth. The gear B meshes with both A and C and is carried on an arm EF which rotates about the centre of A at 18 rpm. If the gear A is fixed, let us determine the speed of gears B and C.

Solution:

Given:

$T_B = 72$;

$T_C = 32$;

$N_A = 18$ rpm (clockwise)

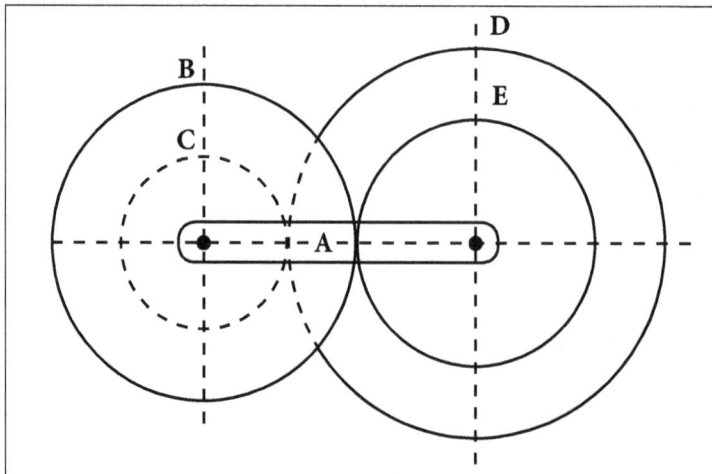

Step No.	Condition of motion	Revolution of elements			
		Arm EF	Gear c	Gear B	Gear A
1.	Arm fixed-gear A rotates through +1 revolution i.e., 1 rev. anticlockwise	0	+1	$-\dfrac{T_C}{T_B}$	$-\dfrac{T_C}{T_B} \times \dfrac{T_B}{T_A} = -\dfrac{T_C}{T_A}$
2.	Arm fixed-gear A rotates through + x revolutions	0	+x	$-x \times \dfrac{T_C}{T_B}$	$-x \times \dfrac{T_C}{T_A}$
3.	Add +y revolution to all elements	+y	+y	+y	+y
4.	Total motion	+y	X+ y	$y - x \times \dfrac{T_C}{T_B}$	$y - x \times \dfrac{T_C}{T_A}$

Speed of gear C,

We know that, the speed of the arm is 18 rpm. Therefore,

$\qquad$ y=18 rpm

The gear A is fixed. Therefore,

$$y - x \times \frac{T_C}{T_A} = 0 \quad \text{or} \quad 18 - x \times \frac{32}{72} = 0$$

$\qquad$ X = 18 x 72/32= 40.5

Speed of gear C = x + y = 40.5 = 18

$\qquad$ = +58.5 r.p.m

$\qquad$ = 58.5 r.p.m in the direction of arm

Speed of gear B,

Let, d_A, d_B and d_c be the pitch circle diameter of gear A,B and C respectively. Therefore,

$$d_B + \frac{d_C}{2} = \frac{d_A}{2} \quad \text{or} \quad 2d_B + d_C = d_A$$

Since, the number of teeth is proportional to their pitch circle diameter. Therefore,

$\qquad$ $2T_B + T_C + T_A$ or $2T_B + 32 = 72$ or $T_B = 20$

Speed of gear B,

$$= y - x \times \frac{T_C}{T_B} = 18 - 40.5 \times \frac{32}{20}$$

= -46.8 rpm

= 46.8 r.p.m in the opposite direction of arm.

2. An epicyclic train of gears is arranged as shown in Figure. Let us how many revolutions does the arm, to which the pinions B and C are attached, make:

- When A makes one revolution clockwise and D makes half a revolution anticlockwise, and

- When A makes one revolution clockwise and D is stationary?

The number of teeth on the gears A and D are 40 and 90 respectively.

Solution:

Given:

$T_A = 40;$

$T_D = 90.$

Let us find the number of teeth on gear B and C(i.e T_B and T_C). Let, d_A, d_B, d_C and d_D be the pitch circle diameter of gears A,B,C and D respectively. Therefore, from the geometry of the figure:

$$d_A + d_B + d_C = d_D \text{ or } d_A + 2 \, d_B = d_D \ldots\ldots (d_B = d_B)$$

Since, the number of teeth is proportional to their pitch circle diameter. Therefore,

$$T_A + 2T_B = T_D \text{ or } 40 + 2T_B = 90$$

$$T_B = 25 \text{ and } T_C = 25 \ldots (T_B = T_C)$$

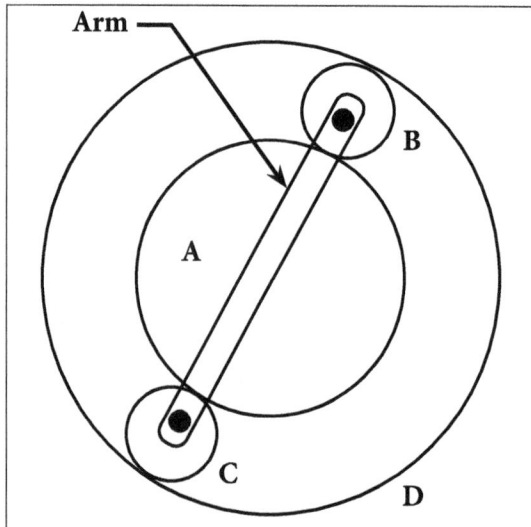

Step No.	Condition of motion	Revolution of elements			
		Arm	Gear A	Compound Gear B-c	Gear D
1.	Arm fixed-gear A rotates through -1 revolution i.e., 1 rev. anticlockwise	0	-1	$+\dfrac{T_A}{T_B}$	$+\dfrac{T_A}{T_B} \times \dfrac{T_B}{T_D} = +\dfrac{T_A}{T_D}$
2.	Arm fixed-gear A rotates through - x revolutions	0	-x	$+X \times \dfrac{T_A}{T_B}$	$+X \times \dfrac{T_A}{T_D}$
3.	Add -y revolution to all elements	-y	-y	-y	-y
4.	Total motion	-y	-X-y	$X \times \dfrac{T_A}{T_B} - y$	$X \times \dfrac{T_A}{T_D} - y$

Speed of arm when A makes 1 revolution clock wise and D makes half revolution anti-clockwise.

Since, the gear A makes 1 revolution clockwise, therefore from the fourth row of the table,

-x - y = -1 or x + y = 1...(i)

Also, the gear D makes half revolution anti-clock wise, therefore,

$$X \times \frac{T_A}{T_D} - y = \frac{1}{2} \quad \text{or} \quad X \times \frac{40}{90} - y = \frac{1}{2}$$

40x - 90y = 45 or x - 2.25y = 1.125... (ii)

From equations (i) and (ii), x = 1.04 and y = -0.04

Speed of arm = -y = -(-0.04) = +0.04

= 0.04 evolution anti-lock wise

Speed of arm when A makes 1 revolution clockwise and D is stationary.

Since the gear A makes 1 revolution clockwise, therefore from the fourth row of the table,

-x-y=-1 or x+y=1...(iii)

Also the gear D is stationary, therefore,

$$X \times \frac{T_A}{T_D} - y = 0 \quad \text{or} \quad X \times \frac{40}{90} - y = 0$$

40x-90y=0 or x-2.25y=0....(iv)

From equation (iii) and (iv),

X=0.692 and y=0.308

Speed of arm= -0.308=-y=0.308 revolution clockwise.

Cams

8.1   Types of Cams

A cam is a rotating machine element which gives reciprocating or oscillating motion to another element known as follower. The cam rotates usually at constant speed and drives the follower whose motion depends upon the shape of the cam.

Almost always the cam is the driver and the follower is the driven. In a cam mechanism, the three essential members are:

- The cam which has a curved surface or straight surface.

- The follower.

- The frame which supports and guides.

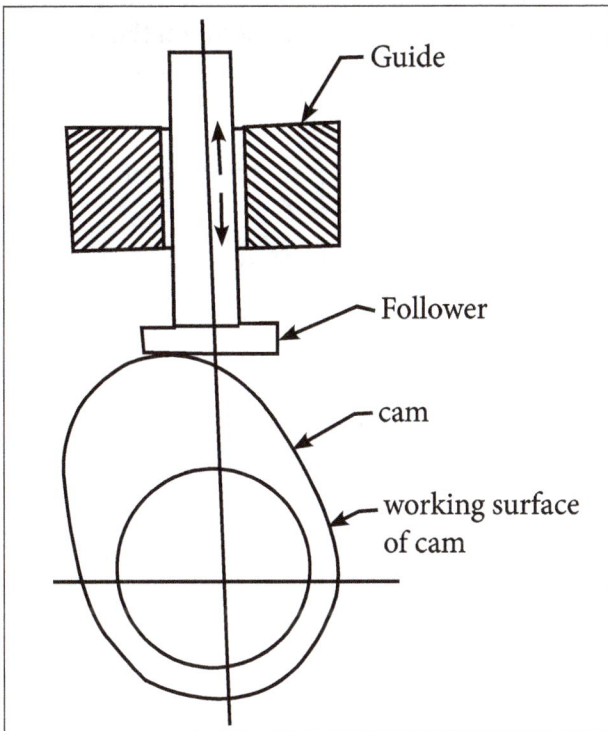

Cam

Cam and Follower

A cam is a mechanical device used to transmit motion to a follower by direct contact. The driver is called the cam and the driven member is called the follower. In a cam follower pair, the cam normally rotates while the follower may translate or oscillate. A familiar example is the camshaft of an automobile engine, where the cams drive the push rods (the followers) to open and close the valves in synchronization with the motion of the pistons.

A cam may be defined as a machine element having a curved outline or a curved groove, which, by its oscillation, rotation or reciprocating motion, gives a predetermined specified motion to another element called the follower.

According to Cam Shape:

- Wedge or Rat Cams.

- Radial or Disc Cams.

- Spiral Cams.

- Cylindrical Cams.

Wedge or rat cams: A wedge cam has a wedge of specified contour. The translation motion of the wedge is imparted to the follower which either reciprocates or oscillates. Generally, a spring is used to maintain contact between the follower and the cam.

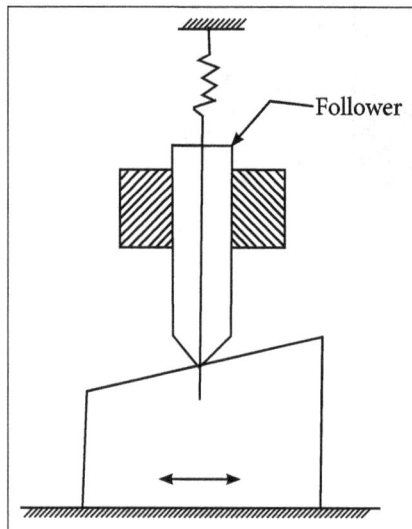

Wedge or Rat Cams.

Radial or disc cams: A cam made out of a plate in such a way that follower moves radially from the centre of rotation is known as plate cam. These cams are also known as disc cams or radial cams because the surface of the cam is so shaped that the follower reciprocates or oscillates in a plane at right angle to the axis of the cam.

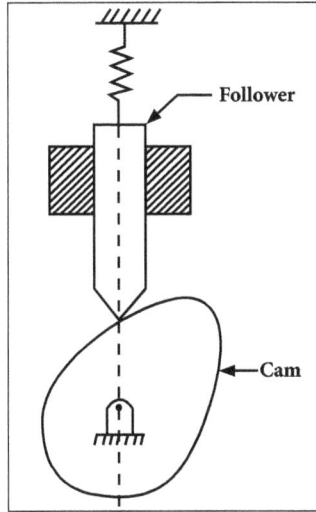

Radial or Disc Cams.

Spiral cams: A circular plate in which spiral groove is cut and a pin gear follower meshes with the teeth cut on spiral groove, as shown in figure below, is called spiral cam. It is also known as face cam. The main limitation of such cam is that it has to reverse the direction to reset the position of the follower.

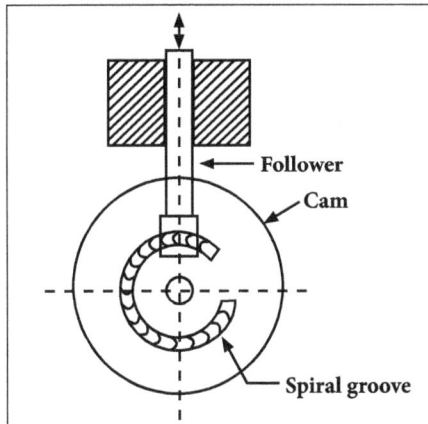

Spiral Cams.

Cylindrical cams: In a cylindrical cam, a circumferential contour is cut on the surface of a cylinder which rotates about its axis. The follower may either oscillate or reciprocate as shown in figures below. These cams are also known as drum or barrel cams.

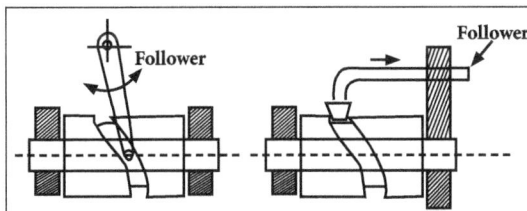

Cylindrical Cams.

8.1.1 Types of Followers

According to the path of motion of follower:

- Radial follower

- Offset follower

Radial follower: When the motion of the follower is along an axis passing through the centre of the cam, it is known as radial followers. Below figures has an example of this type.

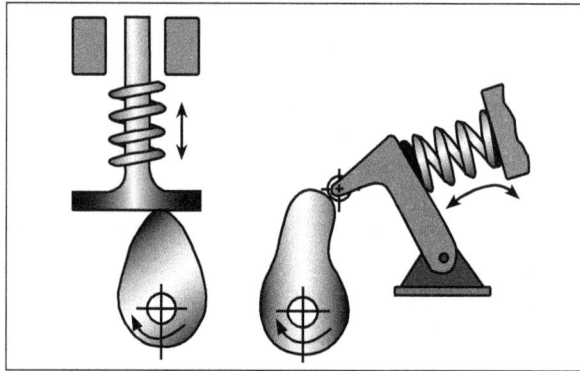

Radial follower.

Offset follower: When the motion of the follower is along an axis away from the axis of the cam centre, it is called off-set follower. Below figures has an example of this type.

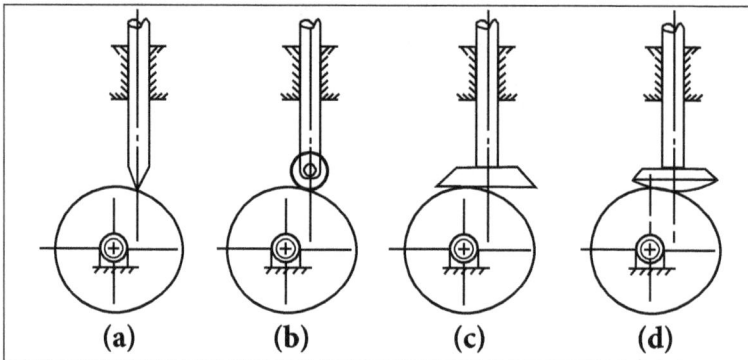

Offset follower.

Cam Follower used in High Speed Engines

For any high speed cam application it is extremely important that not only the displacement and velocity curves, but also the acceleration curve be made continuous for the entire motion cycle. No discontinuities should be allowed at the boundaries of the different sections of cam.

For these reason, cycloidal motion of the follower is used for high speed cams.

8.1.2 Displacement, Velocity and Acceleration Time Curves for Cam Profiles

Simple Harmonic

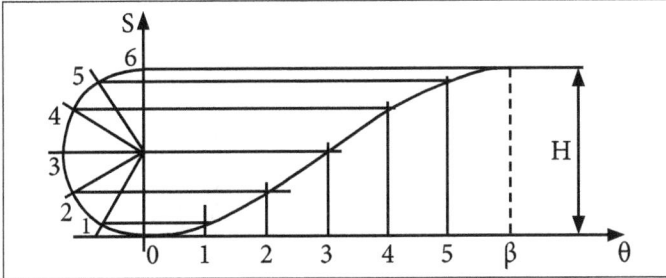

Simple Harmonic Motion.

Simple harmonic motion curve is widely used since it is simple to design. The curve is the projection of a circle about the cam rotation axis as shown in the figure. The equations relating the follower displacement velocity and acceleration to the cam rotation angle are,

$$s = \frac{1}{2}H\left(1 - \cos\left(\frac{\pi\theta}{\beta}\right)\right)$$

$$v = \frac{1}{2}\frac{H\pi\pi}{\beta}\sin\left(\frac{\pi\theta}{\beta}\right)$$

$$a = \frac{1}{2}H\left(\frac{\pi\omega}{\beta}\right)^2\cos\left(\frac{\pi\theta}{\beta}\right)$$

Uniform Velocity

This is where the follower moves at a constant velocity.

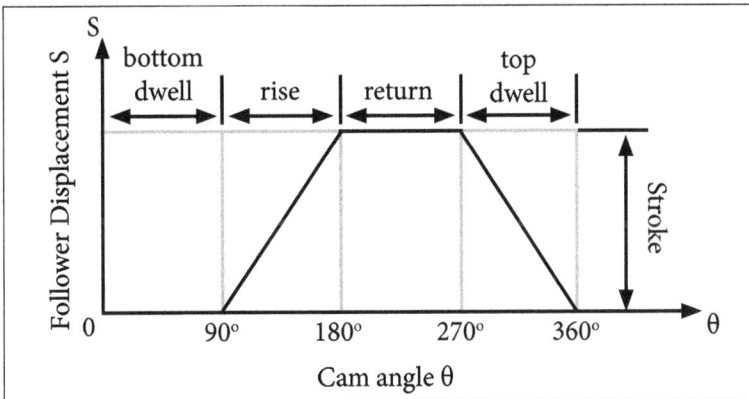

Uniform velocity.

The shock effects inherent from the uniform velocity can be reduced by modifying the motion. The modification is to have the follower undergoing uniform acceleration at the start of the constant velocity interval and uniform deceleration at the end of the constant velocity interval, so that the velocity curve is continuous.

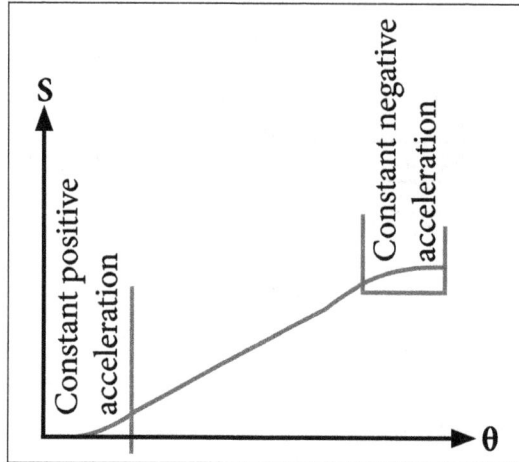

Velocity curve.

Parabolic or Constant Acceleration Motion Curve

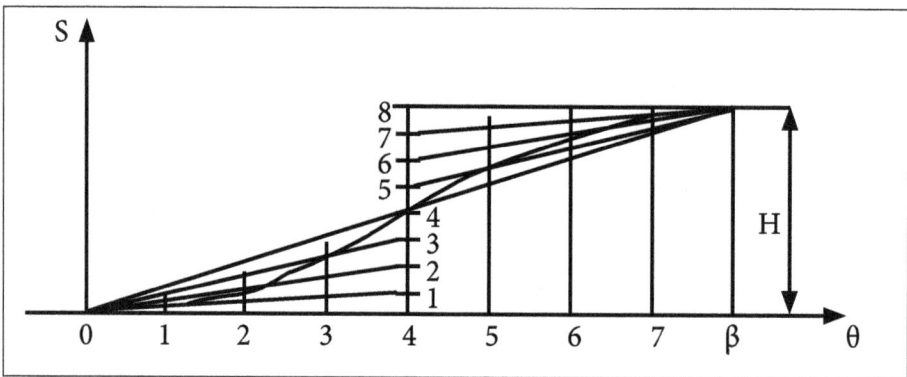

Constant Acceleration Motion Curve.

For the range $0 < \theta < \beta/2$

$$s = 2H\left(\frac{\theta}{\beta}\right)^2$$

$$v = 4H\omega\left(\frac{\theta}{\beta^2}\right)$$

$$a = 4H\left(\frac{\omega}{\beta}\right)$$

For the range $\beta/2 < \theta < \beta$

$$s = H\left[1 - 2\left(1 - \frac{\theta}{\beta}\right)^2\right]$$

$$v = 4H\frac{\omega}{\beta}\left(1 - \frac{\theta}{\beta}\right)$$

$$a = 4H\left(\frac{\omega}{\beta}\right)^2$$

In this case the velocity and accelerations will be finite. There is a constant acceleration for the first half and a constant deceleration in the second half of the cycle. However the third derivative, jerk, will be infinite at the two ends as in the case of simple harmonic motion.

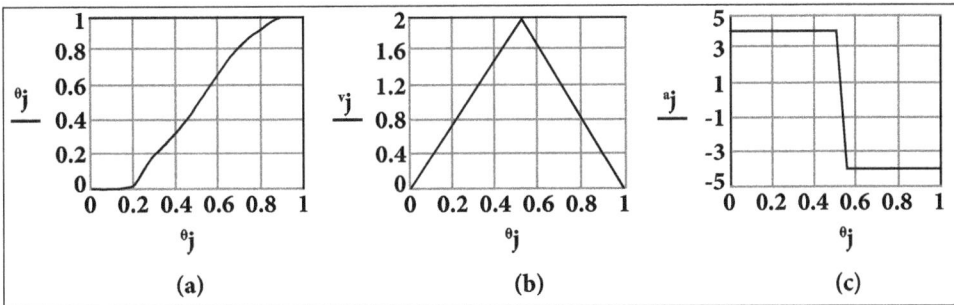

Displacement, velocity and acceleration curves.

Displacement, velocity and acceleration curves are as shown. This motion curve has the lowest possible acceleration.

The following general procedure is adopted for graphical synthesis of cam profile:

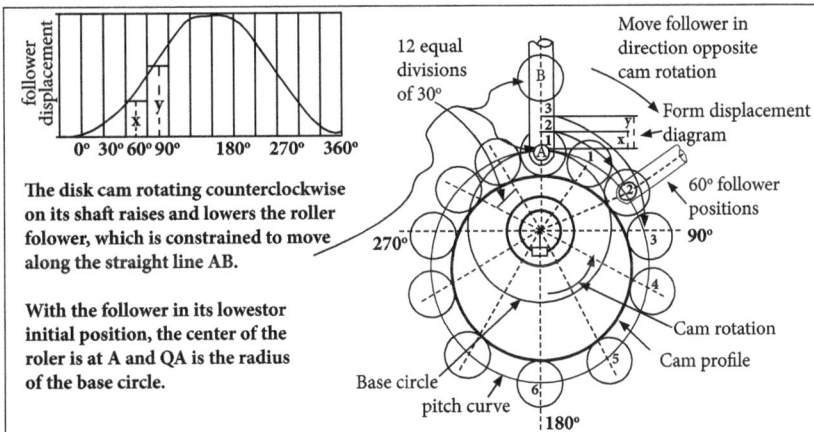

Cam profile.

- Draw the displacement diagram for the given type of follower motion.

- Assume that cam remains stationary and the follower moves around it in the direction opposite to the direction of rotation of the cam.

- In case of knife edge and flat faced followers, draw a base circle of given minimum radius. While in the case of roller follower, draw a pitch circle of radius equal to sum of base circle radius and roller radius.

- Divide its circumference into a number of parts equal to the divisions used in displacement diagram.

- Draw various positions of the follower corresponding to angular position of the cam.

- Draw a smooth curve tangential to the contact surface in different positions of the follower.

8.2   Disc Cam with Reciprocating Follower having Knife-Edge

The disk or plate cam has an irregular contour to impart a specific motion to the follower. The follower moves in a plane perpendicular to the axis of rotation of the camshaft and is held in contact with the cam by springs or gravity.

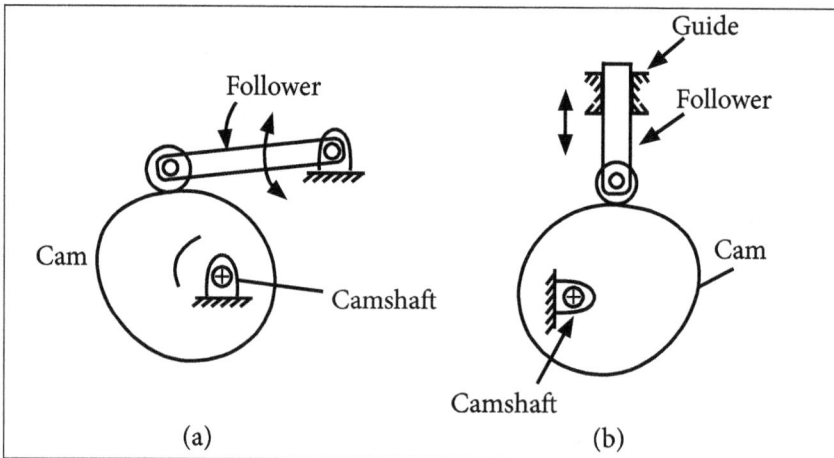

Plate or disk cam.

Reciprocating or Translating Follower

The follower reciprocates in guides as the cam rotates uniformly, it is known as reciprocating or translating follower. The followers as shown in the figure (a) to (c), are all reciprocating or translating followers.

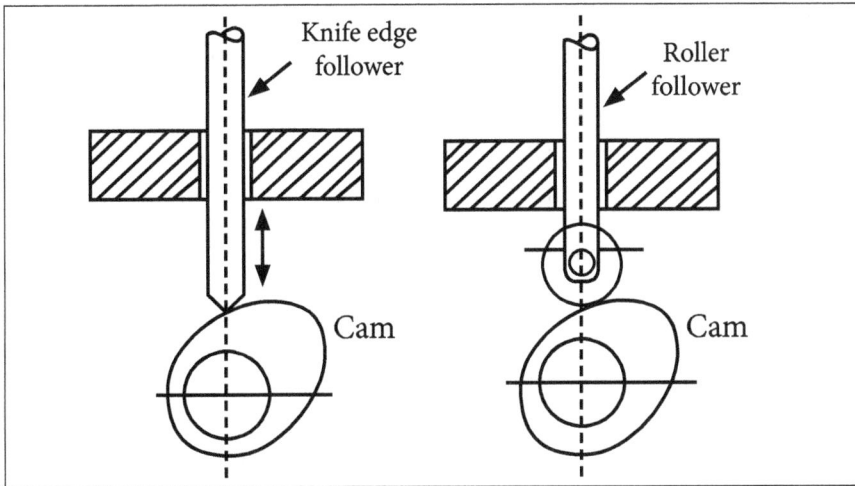

Cam with reciprocating follower.

Knife Edge Follower

The contacting end of the follower has a sharp knife edge, it is called a knife edge follower, as shown in the figure:

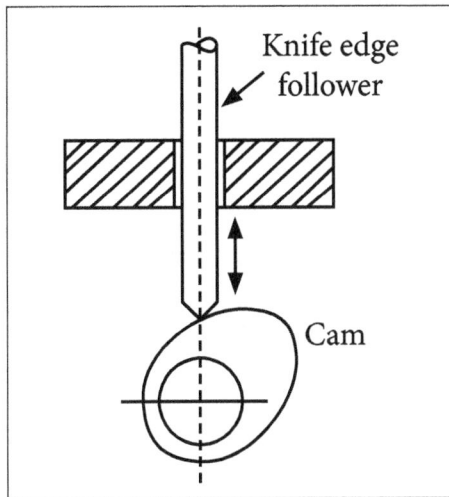

Knife edge follower.

Disc Cam with Reciprocating Follower Having Knife-Edge

This type of cam and follower is shown figure below. The cam is to rotate clockwise while the follower moves radially. The displacement diagram for one revolution of the cam is shown. The diagram has been divided into 12 equal intervals and the cam has been divided into 12 corresponding equal angles.

The distance from the point of the follower in its lowest position to the center of rotation of the cam is the radius of the base circle. Laid off along the follower are the ordinates from each position along the θ axis of the displacement diagram. When the cam rotates two spaces clockwise, the edge of the follower is pushed upward from 0' to 2'.

In order to produce this same relative motion with the cam fixed, the frame is rotated two spaces counterclockwise and the follower is moved outward a distance 0' 2'. Point 2" is located by striking an arc from point 2' using O_2 as a center.

Points 1", 3", 4", etc., around the cam are located in the same manner. A smooth curve through points 0', 1", 2", 3", etc., is then the desired cam profile.

A knife-edge follower is seldom used in practice because the small area of contact results in excessive wear.

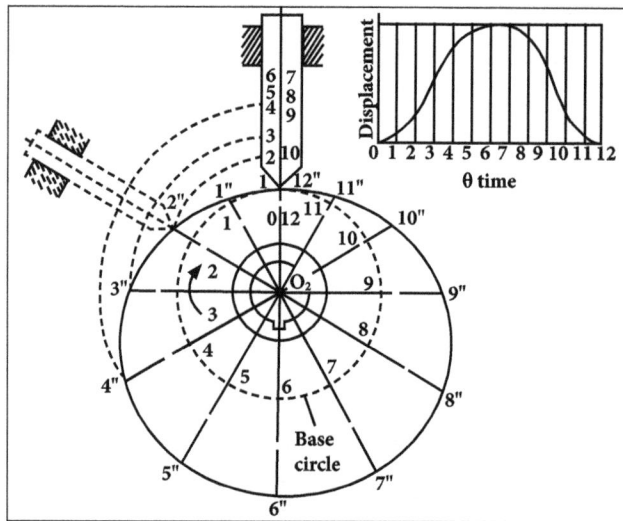

Reciprocating Follower Having Knife-Edge.

8.2.1 Roller and Flat-Face Follower

Roller Follower

The contacting end of the follower is a roller; it is called a roller follower, as shown in Fig;

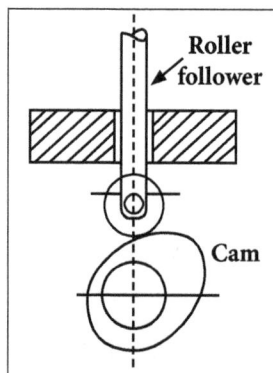

Roller follower.

The cam is to rotate clockwise, while the follower moves according to the scale on its centerline. The base circle passes through the axis of the roller when the follower is in its lowest position. Positions 1", 2", 3", etc., of the roller axis are determined in the same manner as in the previous example.

A smooth curve through these points is the pitch profile. Arcs with radii equal to the roller radius are then struck from these points. A smooth curve tangent to these arcs is the cam profile.

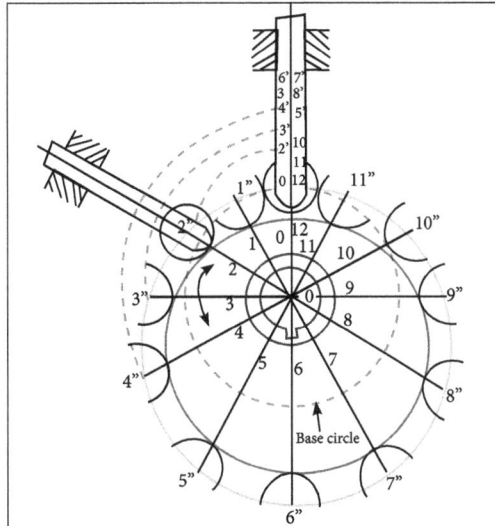

Roller follower.

It is necessary to determine the pitch profile first because the point of contact between the roller and cam does not lie on a radial line through the roller axis unless the follower is in a dwell position. As the cam rotates, the contact point shifts from one side of this line to the other.

Flat-faced or Mushroom Follower

The contacting end of the follower is a perfectly flat face; it is called a flat-faced follower, as shown in the figure:

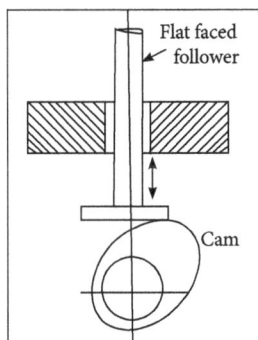

Flat faced or mushroom follower.

The cam is to rotate clockwise and the follower is to move according to the displacements indicated along its centerline. Points 1", 2", 3", etc., are located as in the previous examples. At each of these points a perpendicular is drawn to the radial line on the cam.

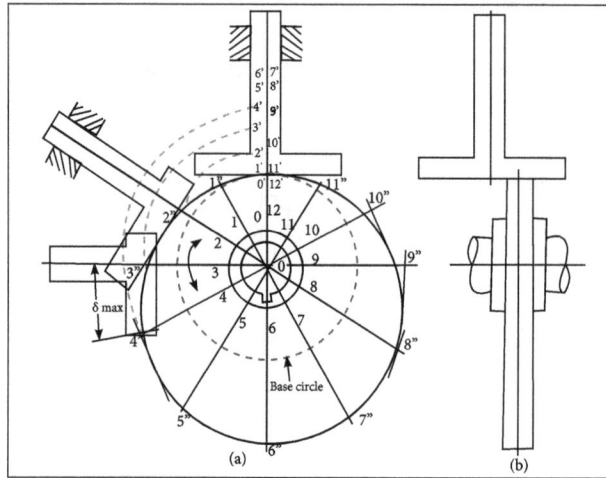

Flat faced or mushroom follower.

The perpendiculars represent the face of the follower as it is rotated around the cam. A smooth curve contacting each of these lines is the cam profile. It can be seen from the figure that the point of contact shifts along the face of the follower. By inspection the maximum deviation of the contact point from the follower centerline is found to occur at phase 3.

This is indicated as the radius of the circular face of the follower is actually made a little larger than what is required to accommodate δm_{ax}. In figure (b), an end view of the cam is shown. The cam is often offset from the follower centerline in order that the follower stem will rotate. This distributes the contact over a larger area on the follower and reduces wear.

Off-set Follower

The motion of the follower is along an axis away from the axis of the cam centre, it is called off-set follower.

The follower, as shown in Fig. below, is an off-set follower.

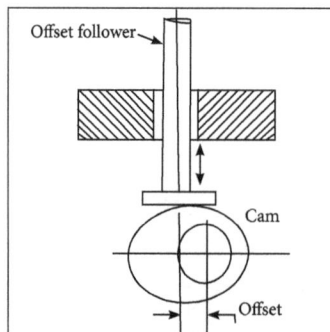

Off-set follower.

Disk Cam with Offset Roller Follower

The distinguishing feature of this kind of cam is that the centerline of the follower does not pass through the center of the camshaft. Sometimes the follower is offset so as to clear another part of the machine. The main reason for this arrangement, however, is that by offsetting the follower, the side thrust on the follower is reduced.

When the offset is to the right, the cam should rotate counterclockwise. When the offset is to the left, the cam should rotate clockwise. These conditions will result in a smaller side thrust on the follower for a given rise in a given angle of cam rotation.

The cam in the figure is to rotate clockwise while the follower moves according to the scale on its centerline. The base circle radius is the distance from the axis of rotation of the cam to the axis of the roller when it is in its lowest position. The perpendicular distance from the center of the cam to an extension of the centerline of the follower determines the radius of the offset circle. This distance is also called the amount of the offset.

The offset circle is divided into 12 equal angles to correspond to the number of equal intervals along the time axis of the displacement diagram. A tangent is then drawn at each position on the offset circle to represent the position of the follower centerline as the cam is held stationary and the frame and follower are rotated counterclockwise about the cam.

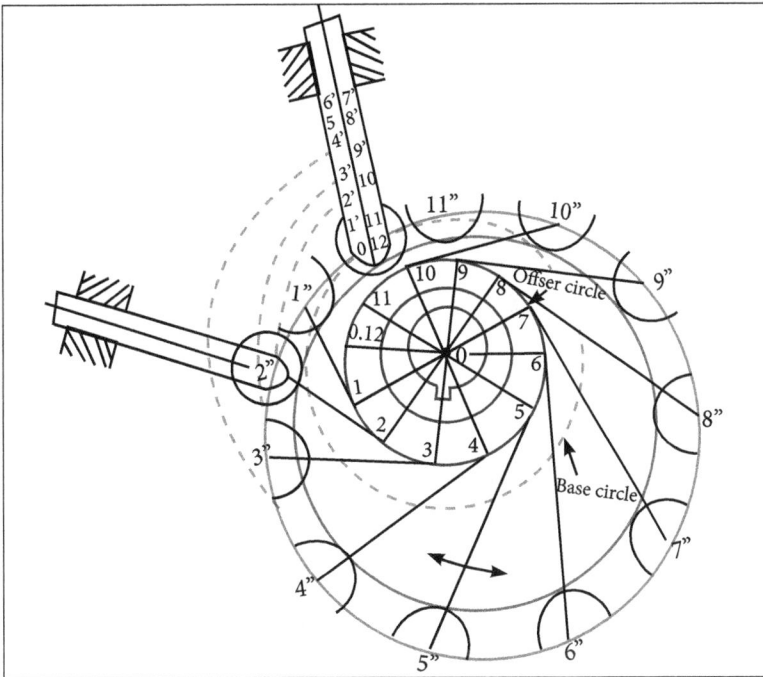

Then with O as a center, arcs are struck from 1', 2', 3', etc., to determine their points of intersection 1", 2", 3", etc., with the tangent lines. A smooth curve through points 1", 2", 3", etc., is the pitch profile. Arcs with radii equal to the roller radius are then struck from these points. A smooth curve tangent to these arcs is the cam profile.

8.2.2 Disc Cam with Oscillating Roller Follower

The uniform rotary motion of the cam is converted into predetermined oscillatory motion of the follower; it is called oscillating or rotating follower. The follower, as shown in the figure below, is an oscillating or rotating follower.

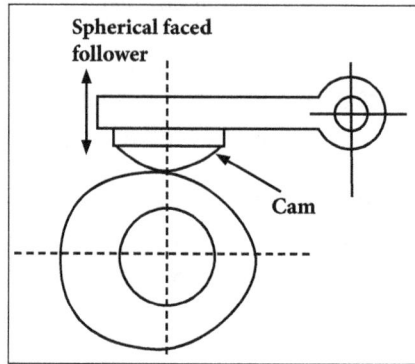

Cam with oscillating roller follower.

The cam is to rotate clockwise and the follower is to oscillate according to the displacements indicated. Point O is the center of the camshaft and point 0' is the lowest position for the face of the follower. The radius of the base circle is the distance from O to 0'. The pivot circle is drawn using the distance from O to the pivot as a radius.

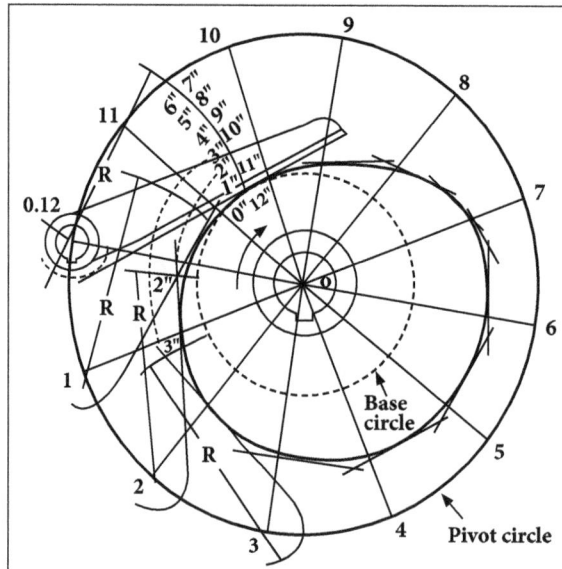

Disc cam with oscillating roller follower.

Next, the positions of the pivot as the follower is rotated around the cam are numbered 1, 2, 3, etc. Then using radius R, arcs are struck using points 1, 2, 3, etc., as centers. Next, with the center of the compass at O, arcs are struck from points 1', 2', 3', etc., to locate points 1", 2", 3", etc. As shown at pivot position 0, the flat face of the follower, when extended, is tangent to a circle of radius r.

As the follower is rotated around the cam, the face of the follower must be tangent to this circle of radius r and must pass through points 1", 2", 3", etc., as shown. A smooth curve contacting each position of the face of the follower is the cam profile.

8.3   Follower Motions Including SHM

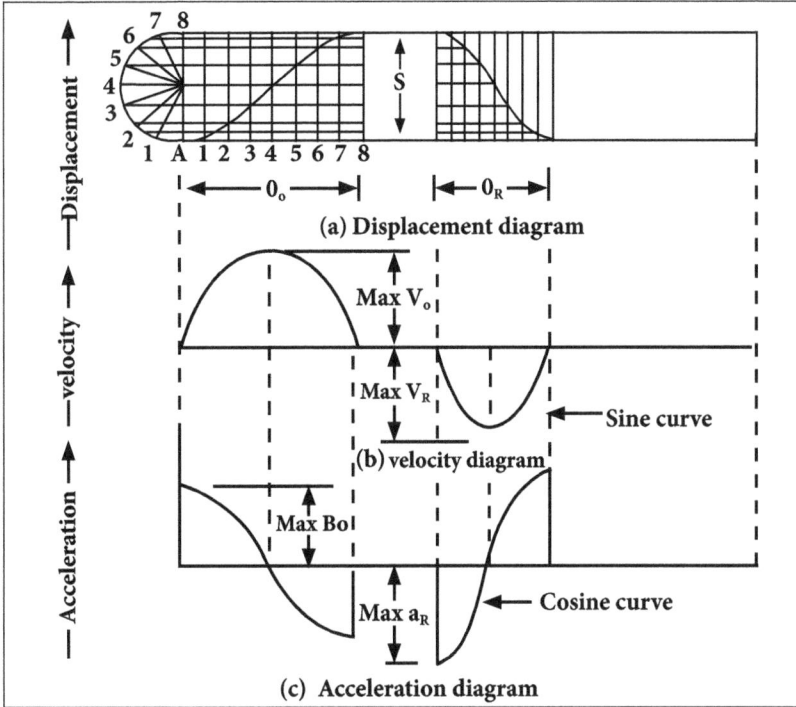

(a) Displacement diagram

(b) velocity diagram

(c) Acceleration diagram

S = Stroke of follower,

θ_o & θ_R = Angular displacement of cam during out & return strokes of follower,

ω = Angular velocity of cam,

Time required for the out stroke of the follower in seconds,

$$t_o = \theta_o / \omega$$

P' executes a SHM as P rotates.

Motion of follower is similar to that of P,

Peripheral speed of the point P,

$$v = \frac{\pi S}{2} \times \frac{1}{t_o} = \frac{\pi S}{2} \times \frac{\omega}{\theta_o}$$

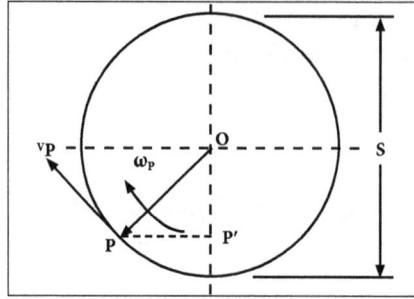

Motion of a Point

Max velocity & Max acceleration of follower on outstroke,

$$v_O = v_P = \frac{\pi S}{2} \times \frac{\omega}{\theta_0} = \frac{\pi \omega . S}{2\theta_0} \qquad a_O = a_P = \frac{\pi^2 \omega^2 . S}{2(\theta_0)^2}$$

Max velocity & Max acceleration of follower on return stroke,

$$v_R = \frac{\pi \omega . S}{2\theta_R} \qquad a_R = \frac{\pi^2 \omega^2 . S}{2(\theta_R)^2}$$

8.3.1 Uniform Velocity

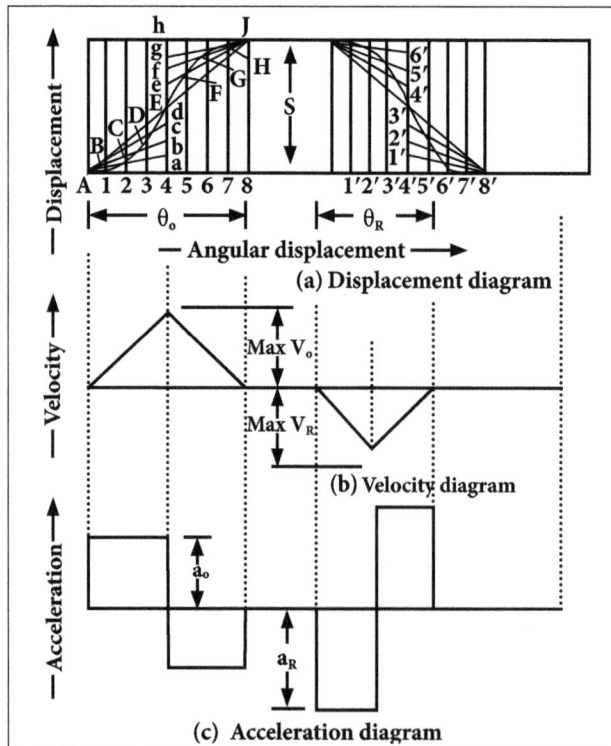

(a) Displacement diagram

(b) Velocity diagram

(c) Acceleration diagram

S = Stroke of follower

θ_o & θ_R = Angular displacement of cam during out & return strokes of follower,

ω = Angular velocity of cam,

Time required for the out stroke of the follower in seconds,

$$t_o = \theta_o / \omega$$

Time required for the out stroke of the follower in seconds,

$$t_R = \theta_R / \omega$$

Mean velocity of follower during outstroke = S/t_o.

Mean velocity of follower during return stroke = S/t_R.

8.3.2 Uniform Acceleration and Retardation

The figure shows the displacement, velocity and acceleration diagram when a follower moves with uniform acceleration and retardation for Dwell-Rise-Dwell cam.

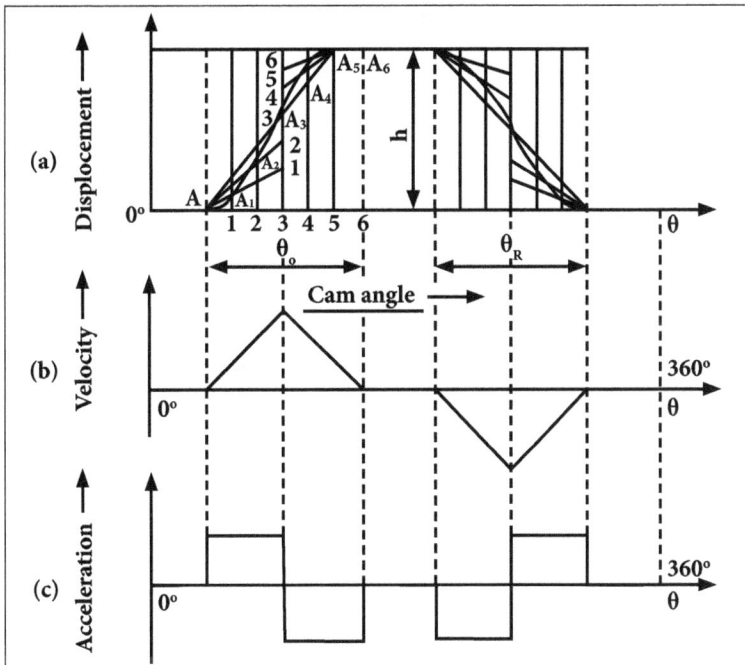

(A) Uniform acceleration and retardation.

Here observe that the displacement diagram consists of a parabolic carve and is constructed as follows:-

- Divide the angular displacement of the cam during outstroke i.e.θ_o into (or during return stroke i.e.θ_R) into even number of equal divisions. Name them as I. 2, 3, ..., 6.

- Divide the lift of the follower (h) into the same number of equal divisions. Name them as 1', 2', 3'...., 6'.

- Join A-1', to intersect the vertical line through 1 at A_1. Similarly obtain the other points A_2, A_3....,A_6 as shown in the figure (a). Now join these points by a smooth curve to obtain the parabolic curve A-A_1-A_2,....., A_6 for the outstroke of the follower.

- In the similar way as discussed above, the displacement diagram for the follower during return stroke may be obtained.

The velocity and acceleration diagrams are shown in the figure (b) & (c) respectively.

Let,

S = Follower displacement (instantaneous).

h= Maximum follower displacement (or lift).

θ= Cam rotation angle (instantaneous).

θ_0= Cam rotation angle for maximum follower displacement, during outstroke.

v = Velocity of the follower.

f= Acceleration of the follower.

Then, general displacement of follower for continuous constant acceleration is given by,

$$S = v_0 t + \frac{1}{2} f t^2$$

Where V_0 = Initial velocity of follower at the start of motion (rise or fall) and is zero in this case,

Therefore,

$$S = \frac{1}{2} f t^2$$

$$f = \frac{2S}{t^2} = \text{constant}$$

As f is constant during the accelerating period, considering the follower at the midway position for outstroke. Referring figure (B), we get,

$$S = \frac{h}{2} \text{ and } t = \frac{\theta_0 / 2}{\omega}$$

$$f = \frac{2(h/2)}{(\theta_o / 2\omega)^2}$$

$$f = \frac{4h\omega^2}{\theta_o^2} \qquad ...(i)$$

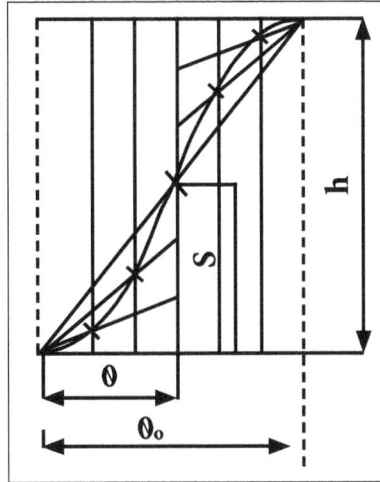

Figure (B)

The velocity is linear during the period and is given by,

$$v = \frac{ds}{dt} = \frac{d}{dt}\left(\frac{1}{2}ft^2\right) = \frac{1}{2} \times 2ft$$

v= ft...(ii)

$$= \left(\frac{4h\omega^2}{\theta_o^2}\right)t$$

$$v = \left(\frac{4h\omega}{\theta_o^2}\right)\theta \qquad ...(iii)$$

Substituting $t = \dfrac{\theta}{\omega}$

The velocity is maximum when θ is maximum or at the follower is at the midway,

i.e. $\theta = \dfrac{\theta_o}{2}$

Therefore, $V_{max} = \dfrac{4h\omega}{\theta_o^2} \cdot \dfrac{\theta_o}{2} = \dfrac{2h\omega}{\theta_o} \qquad ...(iv)$

Equations similar to equations (i), (iii) and (iv) can be obtained, for a return stroke of a follower, by replacing θ_o by θ_R.

Thus from the figure (A), we observe that, there are abrupt changes in the acceleration at the beginning, midway and the end of the follower motion. At midway, an infinite jerk is produced. Hence, this motion to the follower is undesirable for high-speed cams.

Angle of Dwell

Angle of dwell is angle through which the Cam rotates while the follower remains stationary at the highest or the lowest position.

Dwell angle is the angle of rotation of the distributor through which the primary circuit is closed or it is the time where the breaker point is closed. It is designed by the manufacturer according to the number of cylinders.

For example: the dwell angle of 4-cylinders engine may vary between (25-48), but the dwell angle of 6-cylinder engine is between (36 -40). We can conclude that increasing the number of cylinders decreases the dwell angle. In general, the dwell angle may range between (30-60).

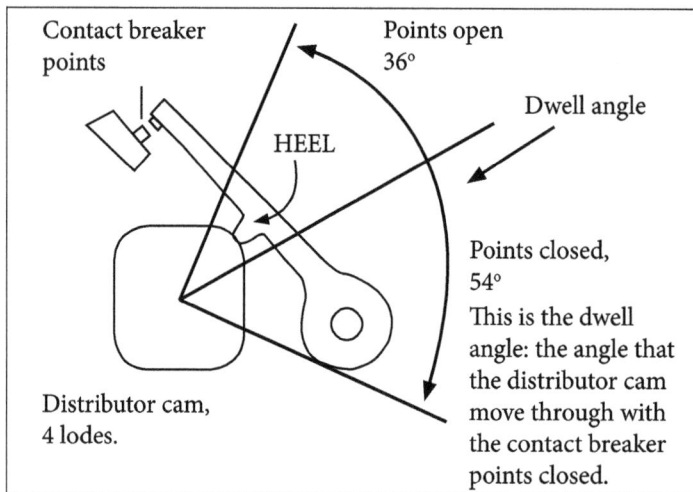

Angle of dwell.

The dwell angle may also defined by the gap. The gap is the distance between the two contact points when they are fully opened. The wider the gap, the smaller the dwell angle.

Gap α_1/Dwell Angle

If the dwell angle is too small, the stored magnetic energy in the coil primary winding will not be sufficient to produce the required high voltage in the secondary circuit. If the dwell angle is too high, the high ignition coil temperature and perhaps coil damage occurs.

Derivatives of Follower Motions

The derivatives of follower motion can be kinematic (with respect to angle) which relate to geometry of the cam system or physical (with respect to time) which relate to the motion of the follower of the cam system.

Straight line-Circular arc motion curve: This curve is an improvement to the linear motion curve. To avoid infinite acceleration at the ends of the rise motion, circles are drawn as shown. Although the acceleration is finite, it will be of a high magnitude.

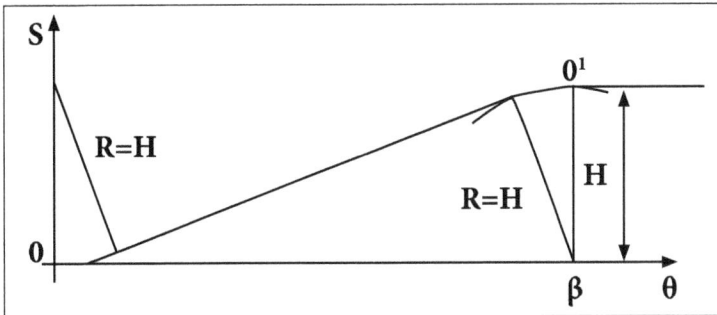

Straight line-Circular arc motion curve.

$$\phi = 2\tan^{-1}\left(\frac{2H^2 + \beta\sqrt{\beta^2 - 3H^2}}{\beta^2 + H^2}\right)$$

$$\theta_1 = H\sin\phi; \quad q_2 = \beta - \theta_1$$

Instead of circular arc the initial and final motions can be simple harmonic or constant acceleration as well as it will be shown in the following example. Straight line motion results in constant velocity. If we are to perform an operation such as cutting during the cam rise, constant velocity is the required motion characteristics.

8.3.3 Cycloidal Motion

The displacement diagram for cycloidal motion is obtained from a cycloid, which is the locus of a point on a circle as the circle rolls on a straight line. In the figure (a), the curve BDE is the displacement diagram for cycloidal motion having a total displacement h while the cam rotates an angle β.

At the right, a circle, whose circumference is h, rolls on the straight line FE. A point on the circle describes the curve FHE, known as a cycloid. As the rolling circle rotates an angle φ, the cam rotates an angle θ. From the figure we note that the displacement s, which is the ordinate to point P on the graph, is equation,

$$s = R\phi - R\sin\phi$$
$$= R(\phi - \sin\phi)$$

Since, the circle makes one revolution for the total rise h,

$$\phi = 2\pi\frac{\theta}{\beta}$$

Also,

$$R = \frac{h}{2\pi}$$

Substituting these last two equations into the equation for s, we obtain,

$$s = \frac{h}{2\pi}\left(2\pi\frac{\theta}{\beta} - \sin 2\pi\frac{\theta}{\beta}\right)$$

$$= h\frac{\theta}{\beta} - \frac{h}{2\pi}\sin 2\pi\frac{\theta}{\beta} \quad \ldots(i)$$

A method for constructing the curve BDE consists of drawing the straight line BE. At any convenient distance to the left on this diagonal the center C of a circle is located. This circle is then divided into a number of equal sectors to correspond to the number of equal segments along the time axis of the diagram. Points on the circle are then projected, as shown, to a vertical line through C.

Next, from the projection of each point on this vertical a line is drawn parallel to line BE in order to obtain the point of intersection with the ordinate of corresponding number on the graph. Six intervals along the time axis were used here. By using more time intervals a more accurate graph can be obtained.

A proof that the construction just explained satisfies equation (i),is as follows. The first term in the equation represents the ordinate to the diagonal BE. The second term represents the length which must then be subtracted in order to obtain the ordinate s.

By differentiation equation (i) and because the angular velocity ω of the cam is constant, the velocity and accelerating equations are found to be,

$$v = \frac{h}{\beta}\omega\left(1 - \cos\frac{2\pi\theta}{\beta}\right)$$

$$A = \frac{2\pi h}{\beta^2} \omega^2 \sin \frac{2\pi\theta}{\beta}$$

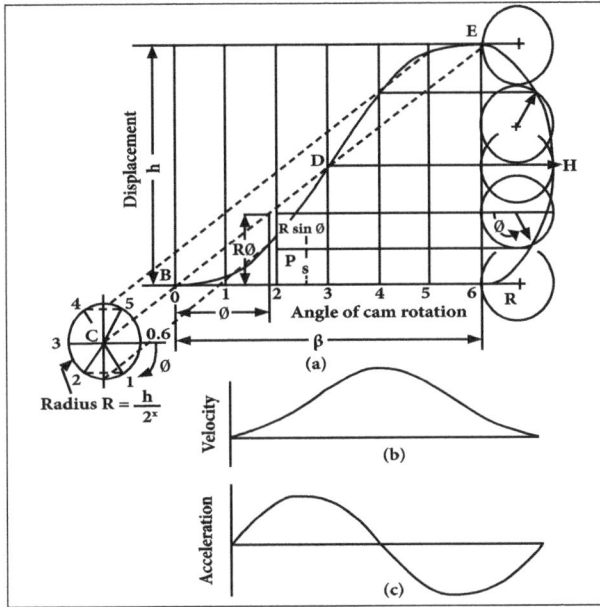

(a)

(b)

(c)

Types of Motion	Displacement	Velocity	Acceleration
Constant acceleration	For $\frac{\theta}{\beta} \leq 0.5$, $s = 2h\frac{\theta^2}{\beta^2}$ For $\frac{\theta}{\beta} \geq 0.5$, $s = h\left[1 - 2\left(1 - \frac{\theta}{\beta}\right)^2\right]$	$\frac{ds}{dt} = \frac{4h\omega\theta}{\beta^2}$ $\frac{ds}{dt} = \frac{4h\omega}{\beta}\left(1 - \frac{\theta}{\beta}\right)$	$\frac{d^2s}{dt^2} = \frac{4h\omega^2}{\beta^2}$ $\frac{d^2s}{dt^2} = \frac{4h\omega^2}{\beta^2}$
Simple harmonic	$s = \frac{h}{2}\left(1 - \cos\frac{\pi\theta}{\beta}\right)$	$\frac{ds}{dt} = \frac{\pi h\omega}{2\beta}\sin\frac{\pi\theta}{\beta}$	$\frac{d^2s}{dt^2} = \frac{\pi^2 h\omega^2}{2\beta^2}\cos\frac{\pi\theta}{\beta}$
Cycloidal	$s = h\left(\frac{\theta}{\beta} - \frac{1}{2\pi}\sin\frac{2\pi\theta}{\beta}\right)$	$\frac{ds}{dt} = \frac{h\omega}{\beta}\left(1 - \cos\frac{2\pi\theta}{\beta}\right)$	$\frac{d^2s}{dt^2} = \frac{2\pi h\omega^2}{2\beta^2}\sin\frac{2\pi\theta}{\beta}$

Problems

1. A cam is designed for a knife edge follower with following data:

- Cam lift = 40 mm during 90° of cam rotation with SHM,

- Dwell for the next 30°,

- During the next 60° of cam rotation, the follower returns to original position with SHM,

- Dwell for the remaining 180°.

Let us draw the profile of the cam when the line of stroke is offset 20 mm from the axis of the cam shaft.

Solution:

Given:

$S = 40 \text{ mm} = 0.04 \text{ m}$.

$\theta_o = 90_o = (\pi/2) \text{ rad} = 1.571 \text{ rad}$.

$\theta_R = 60_o = (\pi/3) \text{ rad} = 1.047 \text{ rad}$.

$N = 240 \text{ rpm}$.

(i) Displacement Diagram

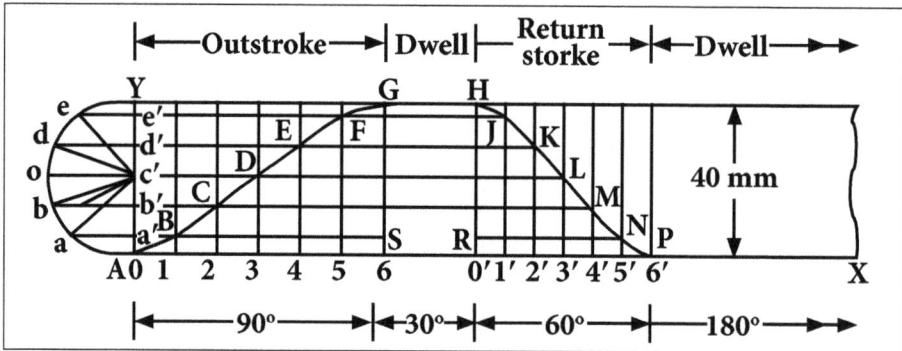

Displacement Diagram.

Displacement diagram is drawn as discussed in the following steps:

- Draw horizontal line AX = 360° to some suitable scale. On this line, mark AS = 90° to represent out stroke; SR = 30° to represent dwell; RP = 60° to represent return stroke and PX = 180° to represent dwell.

- Draw vertical line AY = 40 m to represent the cam lift or stroke of the follower and complete the rectangle as shown in figure (a).

- Divide the angular displacement during out stroke and return stroke into any equal number of even parts (say six) and draw vertical lines through each point.

- Since the follower moves with simple harmonic motion, therefore draw a semicircle with AY as diameter and divide it into six equal parts.

- From points a, b, c ..., etc. draw horizontal lines intersecting the vertical lines drawn through 1, 2, 3, ..., etc. and 0', 1', 2', ... etc. at B, C, D ..., M, N, P.

- Join the points A, B, C..., etc. With a smooth curve as shown in figure. This is the required displacement diagram.

Profile of the cam when the line of stroke of the follower is offset 20 m from the axis of the cam shaft.

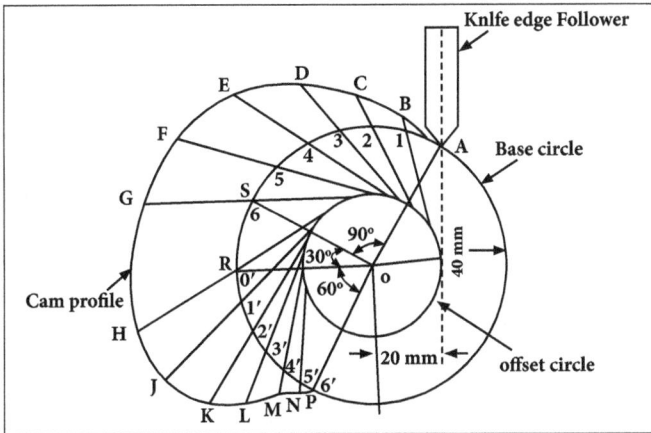

Cam profile.

(ii) Cam profile

Procedure:

- Draw a base circle with radius equal to the Cam lift (40 mm) with O as center.

- Draw the axis of the follower at a distance of 20 mm from the axis of the Cam, which intersects the base circle at A.

- Join AO and draw an offset circle of radius 20 mm with center o.

- From OA mark angle AOS = 90° to represent outstroke, angle SOR = 30° to represent dwell and angle ROP = 60° to represent return stroke.

- Divide the angular displacement during outstroke and return stroke. (i.e, angle AOS and ROP) in the same number of equal even parts as in displacement diagram.

- Now from the points 1,2, 3, ..., etc. and 0', 1', 2', 3', ..., etc. on the base circle, draw tangents to the offset circle and produce these tangents beyond the base circle as shown in figure (b).

- Now set off 1B, 2C, 3D..., etc. and 0'H, 1'J..., etc. from the displacement diagram.

- Joint the points A, B, C..., M, N, P with a smooth curve. The curve AGHPA is the complete profile of the cam.

2. A cam is to give the following motion to a knife-edged follower:

- Outstroke during 60° of cam rotation.

- Dwell for the next 30° of cam rotation.

- Return stroke during next 60° of cam rotation.

- Dwell for the remaining 210° of cam rotation. The stroke of the follower is 40 mm and the minimum radius of the cam is 50 mm.

The follower moves with uniform velocity during both the outstroke and return strokes. Let us Draw the profile of the cam when:

- The axis of the follower passes through the axis of the cam shaft and

- The axis of the follower is offset by 20 mm from the axis of the cam shaft.

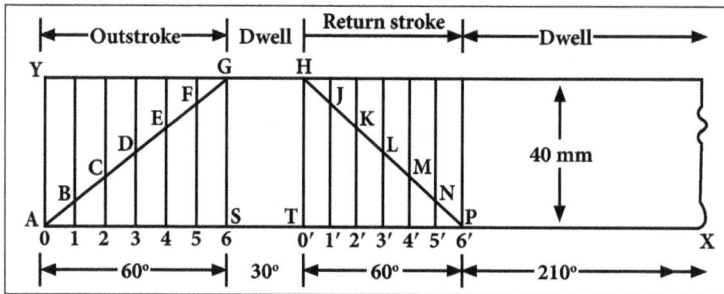

Displacement diagram.

Solution:

The displacement diagram, drawn as discussed in the following steps:

Draw a horizontal line AX = 360° to some suitable scale. On this line, mark AS = 60° to represent outstroke of the follower, ST = 30° to represent dwell, TP = 60° to represent return stroke and PX = 210° to represent dwell.

Draw vertical line AY equal to the stroke of the follower (i.e. 40 mm) and complete the rectangle.

Divide the angular displacement during outstroke and return stroke into any equal number of even parts (say six) and draw vertical lines through each point.

Since the follower moves with uniform velocity during outstroke and return stroke, therefore the displacement diagram consists of straight lines, Join AG and HP.

The complete displacement diagram is shown by AGHPX.

(i) Profile of the cam when the axis of follower passes through the axis of cam shaft:

The profile of the cam when the axis of the follower passes through the axis of the cam shaft, drawn as discussed in the following steps:

Draw a base circle with radius equal to the minimum radius of the cam (i.e. 50 mm) with O as centre.

Since the axis of the follower passes through the axis of the cam shaft, therefore mark trace point A.

From OA, mark angle AOS = 60° to represent outstroke, angle SOT = 30° to represent dwell and angle TOP = 60° to represent return stroke.

Divide the angular displacements during outstroke and return stroke (i.e. Angle AOS and angle TOP) into the same number of equal even parts as in displacement diagram.

Join the points 1, 2, 3 ...etc. and 0′, 1′, 2′, 3′ ... etc. with centre O and produce beyond the base circle.

Now set off 1B, 2C, 3D ... etc. and 0′ H, 1′ J ... etc. from the displacement diagram.

Join the points A, B, C... M, N, P with a smooth curve. The curve AGHPA is the complete profile of the cam.

Notes: The points B, C, D .... L, M, N may also be obtained as follows:

- Mark AY = 40 mm on the axis of the follower and set of Ab, Ac, Ad... etc. equal to the distances 1B, 2C, 3D... etc. as in displacement diagram.

- From the centre of the cam O, draw arcs with radii Ob, Oc, Od etc. The arcs intersect the produced lines O1, O2... etc., at B, C, D ... L, M, N.

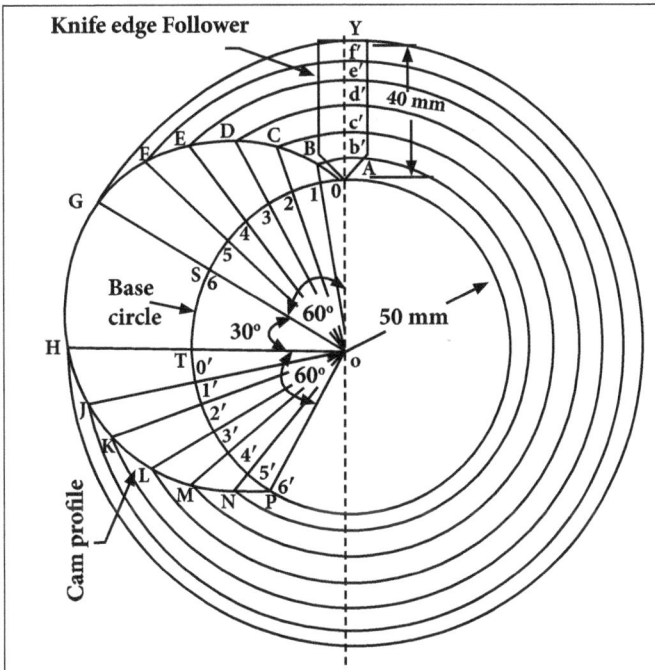

Cam profile.

(ii) Profile of the cam when the axis of the follower is offset by 20 mm from the axis of the cam shaft:

The profile of the cam when the axis of the follower is offset from the axis of the cam shaft drawn as discussed in the following steps:

Draw a base circle with radius equal to the minimum radius of the cam (i.e. 50 mm) with O as centre.

Draw the axis of the follower at a distance of 20 mm from the axis of the cam, which intersects the base circle at A.

Join AO and draw an offset circle of radius 20 mm with centre O.

From OA, mark angle AOS = 60° to represent outstroke, angle SOT = 30° to represent dwell and angle TOP = 60° to represent return stroke.

Divide the angular displacement during outstroke and return stroke (i.e. Angle AOS and angle TOP) into the same number of equal even parts as in displacement diagram.

Now from the points 1, 2, 3 … etc. and 0′,1′,2′,3′… etc. on the base circle, draw tangents to the offset circle and produce these tangents beyond the base circle.

Now set off 1B, 2C, 3D … etc. and 0′ H,1′ J … etc. from the displacement diagram.

Join the points A, B, C …M, N, P with a smooth curve. The curve AGHPA is the complete profile of the cam.

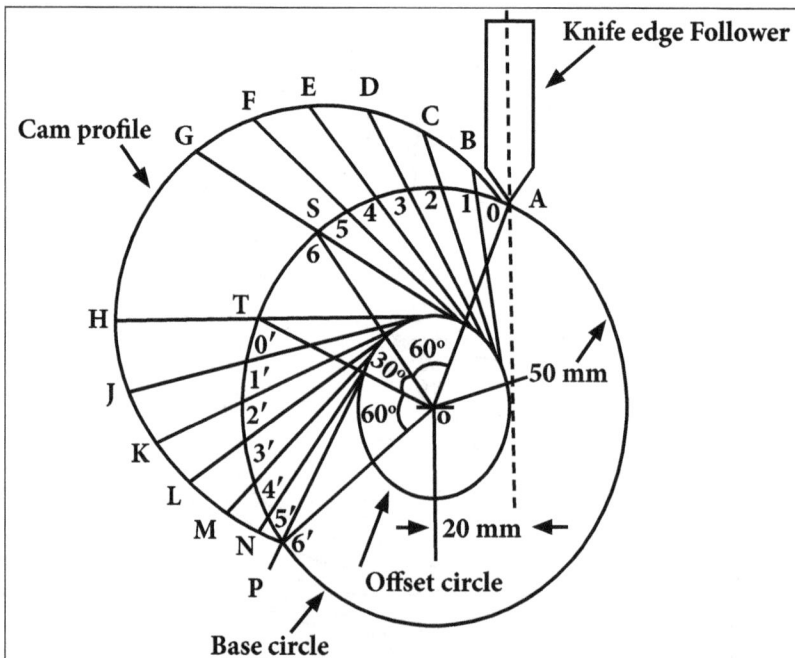

Cam profile.

3. Let us design a cam for operating the exhaust valve of an oil engine. It is required to give equal uniform acceleration and retardation during opening and closing of the valve each of which corresponds to 60° of cam rotation. The valve must remain in the fully open position for 20° of cam rotation. The lift of the valve is 37.5 mm and the least radius of the cam is 40 mm. The follower is provided with a roller of radius 20 mm and its line of stroke passes through the axis of the cam.

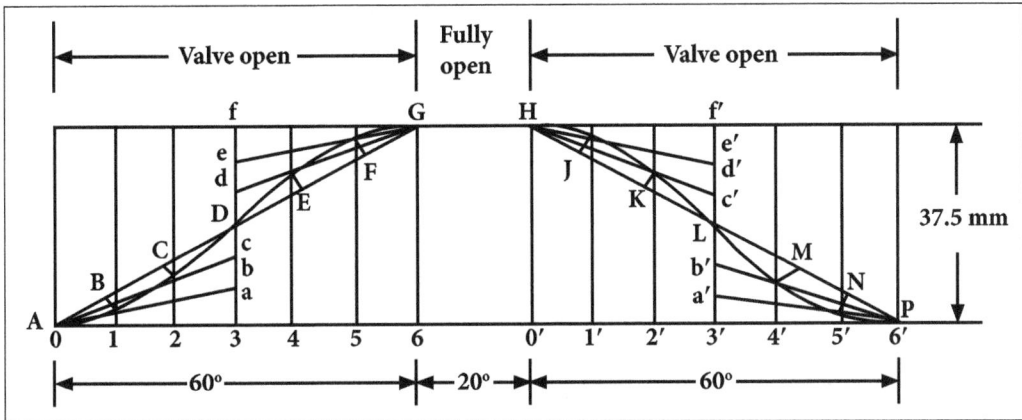

Displacement diagram.

Solution:

The displacement diagram, drawn as discussed in the following steps:

Draw a horizontal line ASTP such that AS represents the angular displacement of the cam during opening (i.e. Out stroke) of the valve (Equal to 60°), to some suitable scale. The line ST represents the dwell period of 20° i.e. the period during which the valve remains fully open and TP represents the angular displacement during closing (i.e. Return stroke) of the valve which is equal to 60°.

Divide AS and TP into any number of equal even parts (say six).

Draw vertical lines through points 0, 1, 2, 3 etc. and equal to lift of the valve i.e. 37.5mm.

Divide the vertical lines 3f and 3′f′into six equal parts as shown by the points a, b, c... And a′, b′, c′... etc.

Since the valve moves with equal uniform acceleration and retardation, therefore the displacement diagram for opening and closing of a valve consists of double parabola.

Complete the displacement diagram.

Now the profile of the cam, with a roller follower when its line of stroke passes through the axis of cam is drawn in the similar way as discussed.

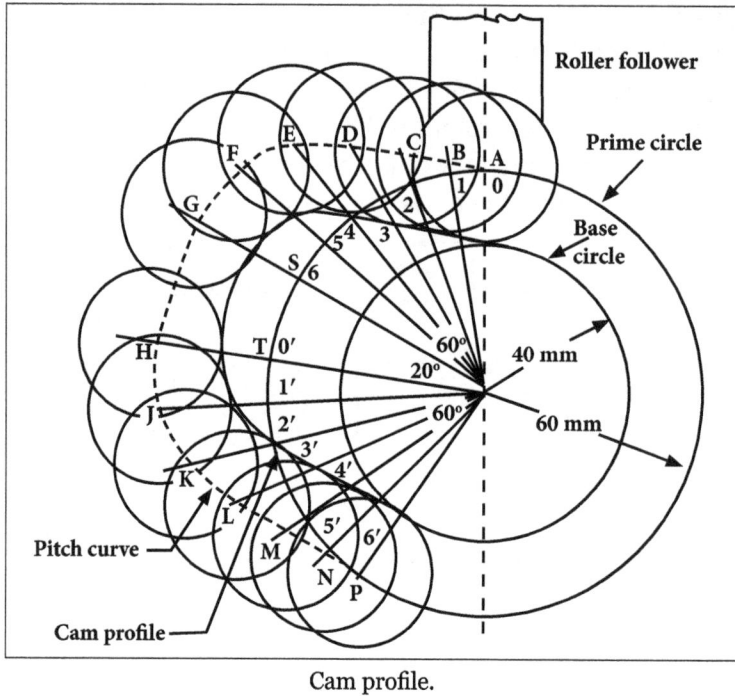

Cam profile.

4. Let us draw a cam profile to drive an oscillating roller follower to the specifications given below: Follower to move outwards through an angular displacement of 20° during the first 120° rotation of the cam, Follower to return to its initial position during next 120° rotation of the cam, Follower to dwell during the next 120° of cam rotation. The distance between pivot centre and roller centre = 120 mm; distance between pivot centre and cam axis = 130 mm; minimum radius of cam = 40 mm; radius of roller = 10 mm; inward and outward strokes take place with simple harmonic motion.

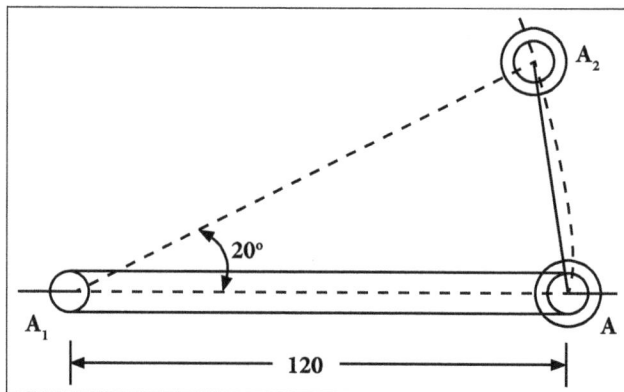

Solution:

We know that the angular displacement of the roller follower = 20° = 20×π/180 = π/9 rad.

Since the distance between the pivot centre and the roller centre (i.e. the radius A_1A) is 120 mm, therefore length of the arc AA_2, as shown in figure along which the displacement of the roller actually takes place.

$$=120 \times \pi/9=41.88 \text{ mm.}$$

($\because$ Length of arc = Radius of arc × Angle subtended by the arc at the centre in radians). Since the angle is very small; therefore length of chord AA_2 is taken equal to the length of arc AA_2.

Thus in order to draw the displacement diagram, we shall take lift of the follower equal to length of chord AA_2 i.e., 41.88 mm.

Displacement diagram.

The outward and inward strokes take place with simple harmonic motion, therefore the displacement diagram, is drawn in the similar way as discussed.

The profile of the cam to drive an oscillating roller follower is drawn as discussed in the following steps:

First of all, draw a base circle with centre O and radius equal to the minimum radius of the cam (i.e. 40 mm).

Draw a prime circle with centre O and radius OA = Minimum radius of cam + radius of roller = 40 + 10 = 50 mm.

Now, locate the pivot centre A_1 such that OA_1 = 130 mm and AA_1 = 120 mm. Draw a pivot circle with centre O and radius OA_1 = 130 mm.

Join OA_1. Draw angle A_1OS = 120° to represent the outward stroke of the follower, angle SOT = 120° to represent the inward stroke of the follower and angle TOA_1 = 120° to represent the dwell.

Divide angles A_1OS and SOT into the same number of equal even parts as in the displacement diagram and mark points 1, 2, 3 . . . 4′, 5′, 6′ on the pivot circle.

Now with points 1, 2, 3 . . . 4′, 5′, 6′ (on the pivot circle) as centre and radius equal to A_1A (i.e. 120 mm) draw circular arcs to intersect the prime circle at points 1, 2, 3 . . . 4′, 5′, 6′, 7.

Set off the distances 1B, 2C, 3D... 4′L, 5′M along the arcs drawn equal to the distances as measured from the displacement diagram.

The curve passing through the points A, B, C....L, M, N is known as pitch curve.

Now draw circles with A, B, C, D....L, M, N as centre and radius equal to the radius of roller.

Join the bottoms of the circles with a smooth curve. This is the required profile of the cam.

Cam profile.

5. The angular velocity of the crank OA is 600 r.p.m. Let us determine the linear velocity of the slider D and the angular velocity of the link BD, when the crank is inclined at an angle of 75° to the vertical. The dimensions of various links are: OA = 28 mm; AB = 44 mm; BC 49 mm ; and BD = 46 mm. The centre distance between the centers of rotation O and C is 65 mm. The path of travel of the slider is 11 mm below the fixed point C. The slider moves along a horizontal path and OC is vertical.

Solution:

Given:

$$N_{AO} = 600 \text{ r.p.m.}$$

$$\omega_{AO} = 2\pi \times 600 / 60 = 62.84 \text{ rad / s}$$

Since, OA = 28 mm = 0.028 m, therefore velocity of A with respect to O or velocity of A (because O is a fixed point),

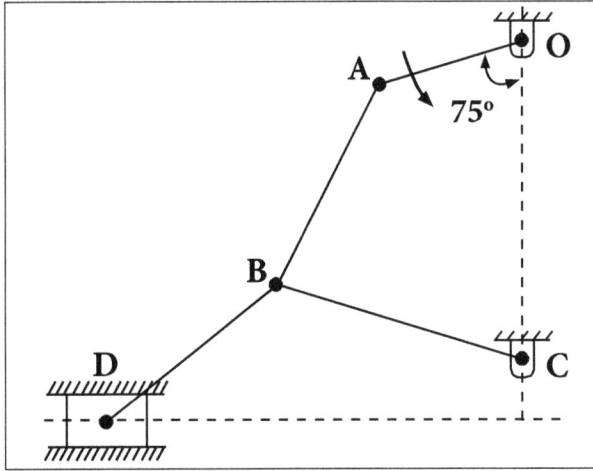

$$v_{AO} = v_A = \omega_{Ao} \times O_A = 0.028 = 1.76 \text{ m/s}$$

Linear velocity of the slider D:

First of all draw the space diagram, to some suitable scale. Now the velocity diagram is drawn as discussed below:

(1). Since the points O and C are fixed, therefore these points are marked as one point, in the velocity diagram. Now from point O, draw vector Oa perpendicular to OA, to some suitable scale, to represent the velocity of A with respect to O or simply velocity of A such that,

Vector Oa = v_{AO} = v_A = 1.76 m/s.

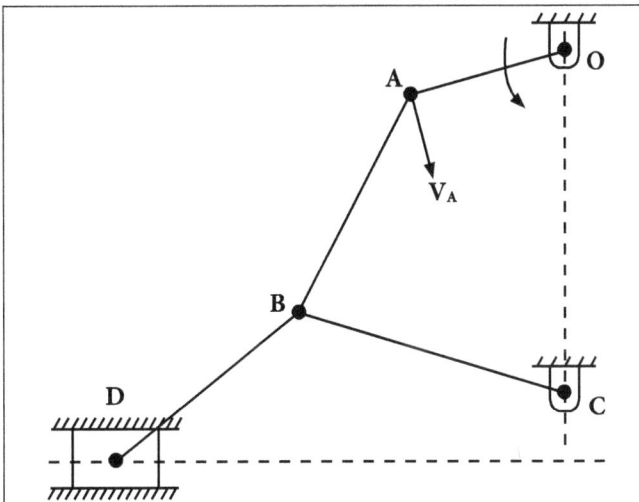

Space diagram.

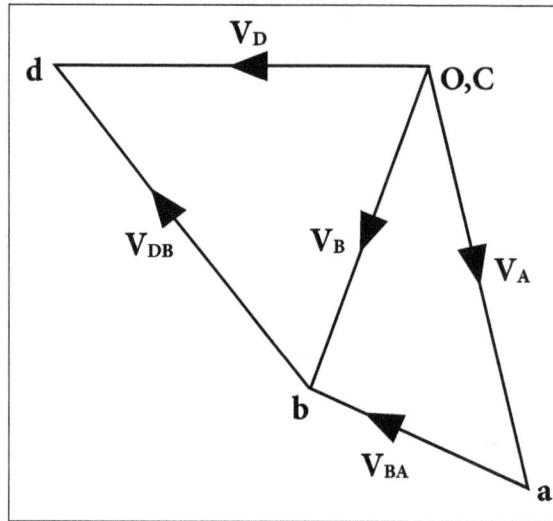

Velocity diagram.

(2). From point a, draw vector ab perpendicular to AB to represent the velocity of B with respect A (i.e. V_{BA}) and from point c, draw vector cb perpendicular to CB to represent the velocity of B with respect to C or simply velocity of B(i.e. V_{BC} or v_B). The vectors ab and cb intersect at b.

(3). From the point b, draw vector bd perpendicular to BD to represent the velocity of D with respect to B(i.e. V_{DB}) and from point O, draw vector Od parallel to the path of motion of the slider D which is horizontal, to represent the velocity of D(i.e. V_D). The vectors bd and Od intersect at d.

By measurement, we find that velocity of the slider D,

v_D = vector Od = 1.6 m/s

Angular velocity of the link BD:

By measurement from velocity diagram, we find that velocity of D with respect to B,

v_{DB} = vector bd = 1.7 m/s

Since, the length of link BD = 46 mm = 0.046 m, therefore angular velocity of the link BD,

$\omega_{BD} = v_{DB} / BD$ = 1.7/0.046 = 36.96 rad/s (clockwise about B)

6. Follower type = Knife edged, in-line; lift = 50mm; base circle radius = 50mm; out stroke with SHM, for 60° cam rotation; dwell for 45° cam rotation; return stroke with SHM, for 90° cam rotation; dwell for the remaining period. Let us determine maximum velocity and acceleration during out stroke and return stroke if the cam rotates at 1000 rpm in clockwise direction.

Displacement diagram:

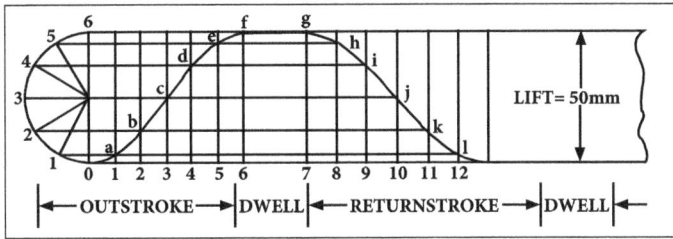

Displacement diagram.

Cam profile:

Construct base circle. Mark points 1,2,3.....in direction opposite to the direction of cam rotation. Transfer points a,b,c.....l from displacement diagram to the cam profile and join them by a smooth free hand curve. This forms the required cam profile.

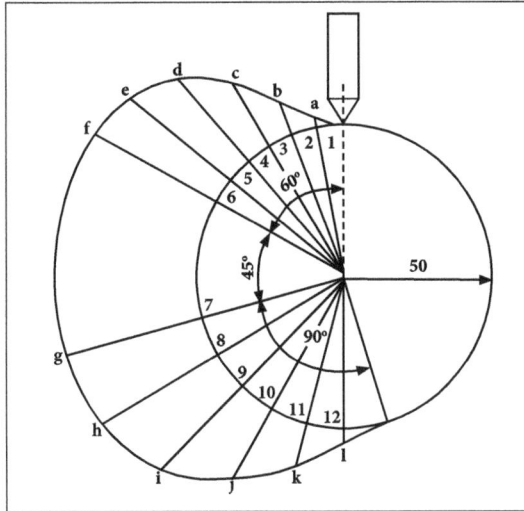

Cam profile.

Calculations:

Angular velocity of cam $= \omega = 2\pi N / 60 = (2\times \pi \times 1000) / 60 = 104.76$ rad / sec

Maximum Velocity of follower during outstroke $= vO_{max} = \pi\omega s / 2\theta_o$

$$= \frac{\pi\times 104.76\times 50}{2\times \pi/3} = 7857 \text{ mm / sec}$$

$= 7.857$ m/s.

Similarly,

Maximum Velocity of follower during return stroke $= vR_{max} = \pi\omega s / 2\theta_r$

$$= \frac{\pi \times 104.76 \times 50}{2 \times \pi / 2} = 5238 \text{ mm/sec}$$

$$= 5.238 \text{ m/s.}$$

Maximum Acceleration during outstroke $= ao_{max} = r\omega^2 p \ (\text{from d3}) = \pi^2 \omega^2 s / 2\theta_o^2$

$$= \frac{\pi^2 \times (104.76)^2 \times 50}{2 \times \left(\pi / 3\right)^2} = 2469297.96 \text{ mm/sec}^2$$

$$= 2469.3 \text{ m/s}_2.$$

Similarly, Maximum Acceleration during return stroke $= ar_{max} = \pi^2 \omega^2 s / 2\theta_r^2$

$$= \frac{\pi^2 \times (104.76)^2 \times 50}{2 \times \left(\pi / 2\right)^2} = 1097465.76 \text{ mm/sec}^2$$

$$= 1097.5 \text{ m/s}_2.$$

Permissions

We would like to thank the editorial team for lending their expertise to make the book truly unique. They have played a crucial role in the development of this book. Without their invaluable contributions this book wouldn't have been possible. They have made vital efforts to compile up to date information on the varied aspects of this subject to make this book a valuable addition to the collection of many professionals and students.

This book was conceptualized with the vision of imparting up-to-date and integrated information in this field. To ensure the same, a matchless editorial board was set up. Every individual on the board went through rigorous rounds of assessment to prove their worth. After which they invested a large part of their time researching and compiling the most relevant data for our readers.

The editorial board has been involved in producing this book since its inception. They have spent rigorous hours researching and exploring the diverse topics which have resulted in the successful publishing of this book. They have passed on their knowledge of decades through this book. To expedite this challenging task, the publisher supported the team at every step. A small team of assistant editors was also appointed to further simplify the editing procedure and attain best results for the readers.

Apart from the editorial board, the designing team has also invested a significant amount of their time in understanding the subject and creating the most relevant covers. They scrutinized every image to scout for the most suitable representation of the subject and create an appropriate cover for the book.

The publishing team has been an ardent support to the editorial, designing and production team. Their endless efforts to recruit the best for this project, has resulted in the accomplishment of this book. They are a veteran in the field of academics and their pool of knowledge is as vast as their experience in printing. Their expertise and guidance has proved useful at every step. Their uncompromising quality standards have made this book an exceptional effort. Their encouragement from time to time has been an inspiration for everyone.

The publisher and the editorial board hope that this book will prove to be a valuable piece of knowledge for students, practitioners and scholars across the globe.

Index